Praise for *Global Warming: Alarmists, Skeptics and Deniers*

"Global Warming: Alarmists, Skeptics and Deniers is a refreshing read on a topic of great societal importance; refreshing because, unlike many books published on this subject, the authors of this work evaluate key predictions and controversies of the global warming debate using logic and science ... Each chapter presents a series of questions and answers that revolve around a central theme. The book is well written and easy to be understood by those with little knowledge or scientific background in the global warming debate."—*Dr. Craig D. Idso, founder and chairman of the board of the Center for the Study of Carbon Dioxide and Global Change.*

"... Robinson leads us systematically through the simple science information that is needed to answer the question, 'Are human carbon dioxide emissions causing dangerous global warming?' And the more surprised you are that the answer to this question is 'no,' then the more you need to read this excellent book."—*Professor Robert M. Carter, Marine Geophysical Laboratory, James Cook University, Australia*

"GLOBAL WARMING: ALARMISTS, SKEPTICS AND DENIERS is an excellent, accessible handbook for those interested in the science of global warming. Skeptics have long maintained that the bulk of the science is on their side. Anyone who reads Dr. Robinson's book with an open mind will find it hard not to agree."—*Iain Murray, author of The Really Inconvenient Truths*

"... an excellent analysis of the problems of climate science from the perspective of a veteran geologist. Particularly useful are Dr. Robinson's discussion of why climate models often fail, and how today's climate changes, when compared to those of the geological past, are clearly seen as well within natural variation. But there is much more science— enough, I'd think, to persuade all but die-hard global-warming believers that a skeptical position is much truer to the climate evidence than alarmism."—*Paul MacRae, author of False Alarm—Global Warming: Facts Versus Fears*

i

GLOBAL WARMING: ALARMISTS, SKEPTICS and DENIERS

A Geoscientist looks at the Science of Climate Change

G Dedrick Robinson Ph.D.
Gene D. Robinson III ESQUIRE

Moonshine Cove Publishing, LLC
150 Willow Point
Abbeville, SC 29620

G Dedrick Robinson Ph.D. recently retired after nearly thirty years as a Professor of Geology at James Madison University in Virginia. He is the author of numerous scientific articles in peer reviewed journals. His background in teaching a wide variety of scientific concepts has uniquely qualified him to discuss difficult scientific concepts in a way that is understandable and interesting to non-scientists. He has studied glaciers in Alaska, volcanic processes in California and Oregon, landslides in North Carolina, flood deposits in Virginia, river processes in Arkansas, zinc deposits in Tennessee, acid mine drainage in Colorado and the conditions which promote the precipitation of manganese oxide coatings on stream alluvium. Dr. Robinson's discussion of climate change and earth history during a public TV program broadcast over a state-wide network was well received. He has spoken at state, regional, national and international geoscience conferences and also has a long record of having successfully taught introductory geology to non-science majors and geology courses to high school earth science teachers.

Gene D. Robinson III Esquire has been practicing law in Virginia since 2007. Prior to establishing his own law practice, he worked as an associate for a law firm in Fairfax, Virginia from 2007 until 2009. He is a 2007 graduate of the University of Florida, Levin College of Law, where he was the Managing Editor of the Florida Journal of International Law. He worked as a summer associate for Fowler White Burnett, P.A., in Miami, Florida and Time Warner, Inc., in New York City. Gene received an Executive Certificate in Financial Planning from Georgetown University in 2003 and a Bachelor of Science in Speech Communication from James Madison University in 1999. Gene is a Certified Financial Planner™ (CFP)® practitioner and worked as a Wealth Management Advisor for United Bank in Vienna, Virginia before attending law school.

Gene is a proud native of the Shenandoah Valley, Virginia where he served as an infantrymen for six years in the Harrisonburg Company of the 29th Light Infantry Division of the Virginia Army National Guard. He is an active member of the Virginia Bar Association, the Fairfax County Bar Association, the Arlington Bar Association, the Northern Virginia Estate Planning Council and the Northern Virginia Financial Planning Association.

"It is a capital mistake to theorize before one has data. Unintentionally one begins to twist facts to suit theories, instead of theories to suit facts."
—Sherlock Holmes in Sir Author Conan Doyle's
A Scandal in Bohemia, 1891

"The tragedy of beautiful theories is that they are often destroyed by ugly facts."

—Thomas Huxley (1825-1895)

"Beware of false knowledge: it is more dangerous than ignorance."
—George Bernard Shaw
Man and Superman, 1903

"By the year 2000 the United Kingdom will be simply a group of impoverished islands inhabited by some 70 million hungry people...If I were a gambler, I would take even money that England will not exist in the year 2000."

—Paul Ehrlich, Speech at British Institute
for Biology, September 1971

CONTENTS

Dedicated to Professor R.H. Carpenter of the University of Georgia. More than a mentor, Bob Carpenter set me on the right path leading to a successful career in science.

PROLOGUE

As a young geology instructor in the 1970s, I informed my students of satellite images showing expanding snow cover in North America compared to previous years. I advised them to remain skeptical of the claims then being made in the popular media that this heralded the beginning of the next ice age. A few years later, sensationalist articles about the coming ice age began to be replaced by others saying the earth was growing dangerously warm and we humans were to blame. Why the continuing exaggerations about climate change I wondered, but duly brought this new scare to the attention of my classes with the same caveat as before. I thought it was just more media hype that would fade as quickly as the recent ice age scare. I was wrong. Instead of fading, global warming alarmism increased.

As a geologist and student of earth history, I knew that climate has always changed. Some of the changes had been disastrous, such as the mountains of ice that moved into mid-latitudes during cold phases of the great Pleistocene ice age depopulating millions of square miles. Mostly, however they were just inconsequential changes of a degree or two. The warming we were then experiencing seemed just the latest in a large number minor undulations of climate, not unusual and not unexpected after the frigid temperatures that lasted several hundred years during the recently ended Little Ice Age. Yet, people were on TV acted like climate change was unusual, something that hadn't happened before. Obviously, they had never taken a historical geology course.

For a while, not many people promoted such a view, but the few who did knew how to get attention. And other people were listening. Little did I know a mindset had taken root and started to grow.

A great many people, particularly liberal politicians and most journalists, act as if climate change is odd, strange, extraordinary. Part of this belief system seems to be that yesterday's climate, the preindustrial climate, was ideal, the best of all possible climates, the way nature intended it to be. It was good because it was natural, and we prospered. But now we have strayed far from the natural way, and with our meddling, have upset the balance. We are the reason climate is changing and since it is not natural, it is bad. Nature gave us a stable climate, an ideal climate, but we messed it up.

I still have trouble coming to grips with this. Do these people not realize that the preindustrial climate was the Little Ice Age chill? Have they not heard of all the crop failures and famine over large areas of Europe? The glaciers moving down into villages, the frozen rivers, the ice-choked harbors? The year without a summer when snow fell in New England during each summer month? The slow starvation of the Viking villages in Greenland?

My answer is that either they do not know these things because they were never exposed to historical geology, or they ignore it in favor of ideology. They either do not know what came before or don't care.

The new climate alarmism that is an offshoot of the green movement, a movement I understand and in some ways, sympathize with. I became a geologist because I love the grandeur of nature and the outdoors and hate the stifling congestion of cities. However these global warming people turned it all around. Staid and unchanging is not nature's way. The most basic thing about the earth, the first thing geologists learn, is that the only thing constant in nature is change. It can be at such a languid pace that, even over one's full lifetime, it's hard to detect, but it can also be catastrophic. Whether hare or tortoise, geologic change can't be stopped, yet it seemed to me that's what global warming alarmists sought.

Earth history clearly teaches that a static earth has never existed. Our planet is one of the most active bodies in the solar system. All kinds of things constantly change, including weather patterns and climate. Still, natural change does not preclude the possibility that human activities might also cause change. Perhaps, I thought, my own predisposition toward natural change was preventing me from impartially assessing the global warming theory. Maybe powerful evidence supported it. With this in mind, I began to study the scientific literature. To my surprise, I found very little direct evidence that humans were influencing climate. Most of what was offered as evidence was based on the predictions of computer climate models, rather than actual observed data or experimental results. It was as if a weather forecast saying sunny skies for the weekend had been elevated to a greater importance than the rain that actually fell. What was going on here?

I eventually realized that a clash of cultures was underway. Geologists hear the deep past, far before human history, speaking to them. Earth history is measured in hundreds of millions and billions of years. Geologists study ongoing geologic processes not to just gain a better understanding of how the earth works today, but also how it worked in the past. Applying the principle of uniformity, we realize that studying present processes helps us learn how the earth worked in the dim recesses of time. This scientific principle, basic to all the sciences, is that scientific laws and processes that operate today also operated in the past. If this were not true, there could be no science because the result of an experiment might turn out differently today than it did last week, and next week, results might differ from either. If nature were chaotic and irregular, no progress in science would be possible, but, except in the subatomic world of quantum mechanics, nature seems to be orderly and regular.

This principle of uniformity also applies in a different way, a very practical way. Studying what happened in the past has proven to be a reliable guide for what is likely to happen again in the future. We use this principle in numerous ways without realizing it. As a child, we learn to speak by applying it. We learn to avoid certain things, such as a flame, because they were dangerous in the past, and likely still are. We've used this principle a long time. To toss it out would be folly.

For global warming alarmists, instead of the past that is living, it's the future. All their dire warnings are based on computer climate model predictions of things that might or might not happen. These are constructed using the best information we have concerning numerous physical, biological and chemical processes thought to control climate. The predictions can only be as good as the information concerning the various controls that are programmed into the computers. Some of this we think we know well, but others are poorly known, or not known at all.

To geologists like me, something vital is missing in this procedure, what we know happened in the past. A vast amount of this sort of data is available, but computer models use none of it to churn out their predictions. Real information, won at great cost and effort, is ignored in favor of predictions. This is not how science is supposed to work. I learned during my years in graduate school that many things in science are important, but above all is the data. We must honor the data, treat it impartially, let it lead us where it will, allow it to illuminate our way toward better theories. This is the only path that will lead to the light of real knowledge, real progress. This new method of science elevates computer predictions above real data. If the data doesn't agree with the computer forecast, then something must be wrong with the data. Better check it again and find out what's wrong.

Along with my discovery that a paucity of evidence supporting global warming was being hyped and stretched almost beyond belief came the realization that studies running counter to the theory were being ignored. Even worse, an entire group of scientists were not being heard, geologists, the very people who have the most knowledge about earth

9

history. The science with the knowledge that should be most helpful in predicting future trends in climate was being ignored. Meteorologists, climatologists, physicists, chemists, biologists, even economists and politicians were making their view known, but where were the geologists?

I finally decided that this oversight needed correcting. This book is the result.

My goal in writing it is to summarize the science of global warming in a way that is understandable to ordinary people. For those desirous of learning more, references to cited peer-reviewed articles are provided. Throughout, I take a geologic point of view, intentionally elevating data above predictions and forecasts, for this is what has made the scientific method so successful in advancing the human condition. I do not discuss the politics of global warming, the economics, the merits of any particular policy direction or any point of view intentionally meant to favor one political party over another, whether liberal or conservative. Plenty of other books cover these topics.

There might be those who say I have failed at one or more of my objectives because they do not like what I say, while others might recommend the book because they do like my viewpoint. This is the unfortunate result of turning an important scientific question into a political piñata. Science and politics mix even worse than oil and water. Each needs to be kept in its cage completely isolated from the other. This is perhaps the most important lesson of the entire global warming controversy. It is in everyone's interest to try to keep it from happening again.

CHAPTER ONE—THE GREENHOUSE EFFECT AND LIFE

Isn't it true greenhouse gases in the atmosphere are causing the earth to heat dangerously?

Well, certain gases in the atmosphere do produce what's popularly known as the greenhouse effect, but the term is misleading because these gases don't work like the glass in a greenhouse and they are not the source of the earth's heat. They're also far more beneficial than dangerous. Before explaining in more detail, some background will be useful.

The nineteenth century Irish physicist John Tyndall is often given the credit for discovering the greenhouse effect. Although not literally true because the existence of gases in the atmosphere producing such an effect was already widely suspected at the time, Tyndall did perform a series of measurements in 1859 which established the existence of such an effect and quantified it. His work determined the capacity of the various gases which compose the atmosphere, including nitrogen, oxygen, carbon dioxide, ozone, water vapor and various minor gases, to absorb infrared energy, or heat.

Recalling high school science, light, the visible portion of the electromagnetic spectrum, lies between the longer wavelength and less energetic infrared, and the more energetic ultraviolet, shorter wavelengths which cause sunburn. The greatest portion of the energy from the sun that reaches earth and heats its surface is in the form of light. The warmed surface is not heated enough for light to produce a visible glow, but it does radiate infrared energy (heat) back toward space. To understand why, think of the familiar analogies red hot and white hot. Red, the longest wavelength part of visible light, comes from a cooler (less energetic) surface than white.

Certain greenhouse gases absorb some of this energy as it's being radiated back toward space, slowing its loss, with the net effect of keeping the surface warmer than it would be otherwise. Tyndall's measurements showed that water vapor is the most important greenhouse gas controlling surface temperature.

Water vapor? Isn't carbon dioxide the most important greenhouse gas?

Not at all, but that is a common misconception due to media hype and ill-informed politicians. Tyndall established this without a doubt, as have many other scientists, so it can stated with certainty. However, the relative importance of water vapor to the total greenhouse effect cannot be stated with the same degree of certainty. This is because the water vapor content of the atmosphere, the humidity in other words, is highly variable from day to day, place to place and constantly changing, ranging from virtually zero in a desert to as much as 4% in certain tropical climates. Physicists who have studied this agree with Tyndall that water vapor accounts for the lion's share of the earth's total greenhouse effect, probably on the order of 90%, but maybe even 95% as S.M. Freidenreich and V. Ramaswamy calculated in 1993.[1] To put it another way, carbon dioxide plus all the minor greenhouse gases combined account for only a few percent of the planet's total greenhouse effect. This is true of the lower portion of the atmosphere, the troposphere where all weather originates and where we live. The relation is reversed in the thin air of the high stratosphere, but this is of very limited importance to the overall greenhouse effect because the temperature in this region is many tens of degrees below zero.

Most people are familiar with the practical importance of humidity as the paramount control of the greenhouse effect, although they might not be aware of it. Consider traveling to Florida during the summer, hot and sticky during the day and only a few degrees cooler at night. Imagine how different it would be hopping on a jet and flying to a desert city such as Las Vegas or Phoenix--as hot or hotter during the day, but rapidly dropping when the sun sets, and getting so cold that a jacket would feel good during the night.

What causes this difference?

Not carbon dioxide, which changes greatly from winter to summer, with less during the northern hemisphere summer because plants use it for growth. Carbon dioxide is pretty well mixed in the atmosphere just like oxygen so there's not much change from place to place. It's the difference in humidity, in other words, changes in the water vapor content of the atmosphere. Where it's higher, such as in Miami, there's a much greater greenhouse effect and temperatures don't fall at night nearly as much as they do in the desert where the air is very dry and the greenhouse effect is diminished.

Another way to look at this phenomenon is to consider how small changes in relative humidity impact the greenhouse effect. As an example, an increase in the relative humidity of only a couple of percent, say from 52% to 54%, hardly even noticeable, changes the greenhouse effect as much as doubling the present carbon dioxide content in the atmosphere.[2]

If this is true, why is carbon dioxide (and maybe methane) the only greenhouse gas the media discusses?

It's most likely not simple oversight, or the fact that water vapor varies so much from place to place. It might be because they don't really understand the greenhouse effect or because of the fact that there's no sensible way to blame humans for variations in humidity, although burning fossil fuels does emit a little water vapor.

Well, it really doesn't matter, does it? An enhanced greenhouse effect will heat the earth, eventually producing a disaster unless we take action to reverse the trend.

There are misconceptions here that should be cleared up regarding greenhouse gases before getting into more specific topics in later chapters, such as how carbon dioxide might be related to global temperatures and what might happen if it increases. It might seem trivial or even a matter of semantics, but the greenhouse effect is a concept important to understand because it is basic to climate change and the global warming theory.

The greenhouse effect, no matter what its cause might be or whether it's increasing, decreasing or static, is not the source of heat for the earth. Except for a modest contribution from the hot interior, the sun is the earth's *only* source of heat. All the greenhouse effect does is to slow the loss of the sun's energy back into space; it does not stop or prevent it. To use the example of Florida again, the temperature doesn't drop much at night during the summer, but it does drop. Energy from the heated surface is lost. What would happen if the sun didn't rise the following morning and on subsequent mornings? Heat loss would continue until the surface cooled far below freezing.

In this respect, the greenhouse effect acts something like insulation in a house. A furnace is capable of producing a certain amount of energy. Depending on the outside temperature, it can warm a house by a certain amount. The reason it cannot warm it beyond that amount is that the heat energy is constantly being lost through the windows, walls, floor, roof, not to mention the cracks and crevices, to the outside environment. If insulation is added, the heat loss is slowed and the furnace can heat the house more than it could before, but heat loss still occurs. It is not stopped completely.

This is an overly simple analogy that is nevertheless useful for understanding the greenhouse effect of the earth, as long as one is aware of the model's limitations. In fact, the earth's atmosphere does not actually work like insulation, a blanket, the glass in a greenhouse (or car) or any other physical barrier to heat flow. Greenhouse gases do not form a physical barrier preventing heat loss.

How does the atmosphere's greenhouse effect differ from these?

Greenhouse gases absorb some of the energy from the sun as it passes down through the atmosphere and more as it is reradiated at a longer wavelength back out into space. A constant exchange of energy occurs with the net effect that greenhouse gases slow heat loss to space due to radiation, but they do not form a physical barrier preventing convectional heat exchange, which is the main way heat exchange occurs in the atmosphere. The glass in a greenhouse does both.

This is important because convectional circulation is by far the most important mechanism for exchanging heat in the atmosphere. Hot air from the equatorial regions is carried toward the poles while cold arctic and Antarctic air moves in the opposite direction, giving rise to wind and weather. According to the calculations of atmospheric physicists, such convectional circulation causes the earth's temperature to be more than 100 degrees F cooler than it would be without convection, which is good unless a toasty 171 degree F day appeals. That's roughly what the earth's average temperature would be without convectional mixing of hot and cold air.[3]

Well, maybe greenhouse warming won't stop convection from operating, but it is still has the potential to cause dangerous warming. Isn't that correct?

Oh, the much maligned greenhouse effect--it never gets any credit. Everything from heat waves and floods to blizzards and drought gets laid at its doorstep. It should wise up and hire a good publicity agent.

To try to set the record straight--the fact that the atmosphere of the earth produces a greenhouse effect is NOT a bad thing. In fact, it's a very good thing. Without the greenhouse effect, there would be no life on the earth.

That's right. The greenhouse effect makes life possible on the earth. Of course, it's not the only factor contributing to the existence of life on this planet. There are many others, but it is one of them.

The reason is simple--temperature. Just as insulating a house helps make it warmer in the winter, the atmosphere's greenhouse effect slows heat loss from the surface and troposphere making the earth warmer than it would be otherwise.

How much warmer?

Estimates for the average temperature of an earth with an atmosphere and clouds but no greenhouse effect vary from a couple of degrees above zero F to a few degrees below zero, but all of them are far below freezing, even at the equator. There would be no liquid water and the earth would be a giant ball of ice, including the oceans, making life impossible.

It is interesting that the earth would be considerably warmer than this, just below freezing in fact, if it had no atmosphere at all. The reason is that clouds reflect a lot of the sun's incoming energy sending it back out into space before it can warm the surface. Clouds, of course, imply an atmosphere of some sort. The calculations from physics for determining earth's temperature without clouds or an atmosphere are shown at junkscience.com.[4]

Another way to look at this is that clouds increase the planet's *albedo*, the proportion of light striking the planet's surface that is reflected back into space. Measured albedo is always between 0 (a surface existing only in theory that reflects no light at all) to 1 (a theoretical perfect reflector). Other factors being equal, the higher the albedo, the brighter an object appears. The albedo of the earth is high, 0.37 compared to 0.13 for the moon and 0.15 for Mars, but Venus is considerably higher at 0.65 partially accounting for why the so-called evening star and morning star, depending on time of the year, is so much brighter in the sky than Mars.

Instead of either of these freezing scenarios, our planet's atmosphere has a lot of water vapor so we have a healthy greenhouse effect. As a result, the average temperature of the earth is well above freezing which is why we have so much liquid water. In fact, the surface is nearly three quarters ocean, with just over one quarter land. There were times in past geologic periods when water covered an even greater proportion of the surface and times when there was more land than today. We are currently somewhere between such extremes.

There is another important aspect of the greenhouse effect that never gets mentioned during media discussions of climate change and global warming. Perhaps it's of lesser importance because life *might* be possible without it, but if so, such life would face far greater challenges than is currently the case in terms of being able to quickly adapt to huge changes in temperature in a short period of time. This is precisely what would be required in order for living things to survive on a planet that lacked a greenhouse effect.

For an example of this, we only have to look at our nearest neighbor in space, the moon or Luna as it was known in Roman times. It's a world composed of rock that in some ways resembles the earth, but it's far smaller so its gravity is much less. In fact, its gravity is so low that it was unable to retain its atmosphere if it ever had one. Conveniently close to us is a world lacking a greenhouse effect that is the same distance from the sun as the earth, so each square meter receives the same amount of energy. What sort of temperature exists there? Close to the average of a degree or so below freezing mentioned earlier for the earth if it had no atmosphere?

That's not a bad estimate, but it only goes to show how misleading using "average" in reference to temperature can be. At morning, right before the sun begins to peep above the horizon, the surface temperature on the moon is far colder than anything remotely approached on the earth, hundreds of degrees F below zero, but the temp begins to rise as the sun climbs higher in the sky until the it's well above the 212 degrees F required to boil water on earth at sea level. The average daytime temperature is 225 degrees F while the night averages -243 degrees F, more than a hundred degrees colder than ever measured on earth. It is doubtful that any life from the earth could adapt to such extremes of heat and cold.

Of course the earth doesn't have temperature extremes like that, but there is a considerable variation between the poles and the equatorial regions. What is the average temperature on the earth?

That seems like such a simple question to answer, but as with so many questions dealing with climate, it turns out to be a surprisingly difficult problem. In fact, the best we can do is to make estimates that agree fairly well with each other.

First, there's the problem of defining what an answer to this question means. Is it enough to just take the average annual temperature for all weather stations and calculate a grand average? That would be biased to areas with people because that's where weather stations are located. What about all the wilderness areas, the oceans and Antarctica? How do we include those? And is one year's data enough? If not, how long a period do we need? And if climate is changing, as it always does, what are we really measuring?

These difficulties have been recognized since at least 1878,[5] but have not stopped scientists from tackling the problem. Older sources typically gave a number on the order of 15 to 18 degrees C (57 - 64 degrees F), but modern sources place it closer to 14 degrees C, roughly 58 degrees F.

Even though a single number cannot be given with certainty in answer to this problem, the rather small range that has been determined makes it clear that our atmospheric greenhouse effect has led to a world much more suitable as a home to life than would otherwise exist. This basic fact is one that should not be overlooked in any discussion of climate and the greenhouse effect, but the media rarely mentions it. It is not unusual for the term "greenhouse effect" to be used interchangeably with climate change, but the two actually refer to completely different phenomenon. An "enhanced greenhouse effect" is closer to how global warming is often used, but even that term can be a bit slippery. A good piece of advice for anyone trying to understand the climate change controversy is to be precise in the use of terms and to be on guard for misused terminology, whether intentional or otherwise.

SUMMARY

Scientists have known since the nineteenth century that the earth's atmosphere produces a greenhouse effect, that is, certain gases absorb some of the longer wavelength radiation that travels back toward space from the sun-heated surface of the earth, thereby helping to retain some of the heat. Scientists of that time compared this to the way glass in a greenhouse traps heat from the sun. This is a limited analogy because greenhouse gases do not form a physical barrier preventing both radiation and convection like the glass in a greenhouse does. The effect of absorbing some of the radiation is to slow the loss of heat rather than preventing it, setting up an exchange of heat between the surface and the lower atmosphere.

Irish scientist John Tyndall in 1859 was the first to quantify the ability of various atmospheric gases to contribute to the greenhouse effect. He found that water vapor is far more important in this regard than all other greenhouse gases combined. Modern studies suggest that as much as 95% of the total greenhouse effect is due to water vapor in the atmosphere alone. An increase in the relative humidity of the air by less than two percent increases the greenhouse effect as much as doubling carbon dioxide from the current level.

Although global warming alarmists and the media often use the term "greenhouse effect" in a negative sense as a synonym for global warming, the two are not the same. The fact that the earth's atmosphere produces a greenhouse effect is one factor that allows life to exist on this planet. It keeps surface temperatures from dropping precipitously at night and from rapidly heating to intolerable levels during the day through the constant ongoing exchange of energy between the surface and the atmosphere. By slowing the loss of heat energy into space, moderate temperatures suitable for liquid water are maintained over most of the planet's surface.

If the atmosphere did not produce a greenhouse effect, all oceans, rivers and lakes would be solid ice with an average global temperature near zero degrees F or below and life as we know could never have developed. Instead, we live on a greenhouse-warmed world with a relative balmy surface temperature averaging somewhere in the neighborhood of 58 degrees F (14 degrees C).

NOTES AND SOURCES

(1) *Journal of Geophysical Research*, Vol. 98, p. 7255, 1993, S.M. Freidenreich and V. Ramaswamy, "Solar Radiation Adsorption by Carbon Dioxide, Overlap with Water, and a Parameterization for General Circulation Models."
(2) Many sources point out the importance of water vapor in relation to carbon dioxide. One of the most accessible is here:
http://www.junkscience.com/Greenhouse/
(3) *Ibid*
(4) *Ibid*
(5) *Nature*, Vol. 17, p. 202, 1878, D. Traill, "Average Annual Temperature at earth's Surface."

CHAPTER TWO—CARBON DIOXIDE, GREENHOUSE GASES AND CLIMATE

Is it correct that the EPA considers carbon dioxide to be a dangerous pollutant and recommends that it be controlled and regulated just like any other atmospheric pollutant?

With the election of Barack Obama in 2008, this was no surprise. In March 2009, the EPA sent a finding to the White House stating that carbon dioxide and five other atmospheric gases are pollutants that endanger public health. *Time Magazine* announced this on their web site with the headline, EPA CALLS CO2 A DANGER--AT LAST.[1] This is certainly an interesting development, particularly since the EPA admits that carbon dioxide is emitted by natural processes. It makes one wonder just what definition for the word "pollutant" they might be using, for if they are right, then each and every person on the earth is producing pollution with each exhaled breath. All other air breathing and carbon dioxide exhaling animals also fall into this category. We are all big-time polluters and are not about to stop.

Although that might be an interesting observation, it doesn't address whether too much carbon dioxide is dangerous and if it is correct to call it a pollutant.

Instead of slogging through a discussion of semantics that leads to no useful destination, a wiser course might be to consider some basic information about carbon dioxide in order to better understand how it relates to climate. A good place to start is the history of the atmosphere and the carbon cycle.

Contrary to media portrayals and popular belief, from a geologic standpoint, the amount of carbon dioxide currently in the earth's atmosphere is unusually low, not high. It is common knowledge among geologists that the atmosphere contained far more carbon dioxide for much of the geologic past.

Geoscientists who have studied past atmospheric composition think carbon dioxide was much higher in the earth's earliest atmosphere, perhaps as much as 10%. That's over 250 times more carbon dioxide than today's level. This early atmosphere was strongly reducing, quite unlike today's oxidizing or oxygen-rich atmosphere.[2] In simplest terms, this means that if it were possible to climb into a time machine and travel back 3.8 billion years ago, the air would be breathable because it lacked oxygen.

No oxygen. Then where did it come from?

Oxygen started to build up in the atmosphere about 2 billion years ago as a waste product from the utilization of carbon dioxide in one of the most important chemical processes ever initiated on the earth, photosynthesis. Yes, photosynthesis, that hoary old favorite of eighth-grade science everyone had to memorize in grade school, is the reason the atmosphere of the earth has more than 20% free molecular oxygen making it unique among all known planets. Oxygen is a product of life and a telltale sign or signature. Free molecular oxygen in a planet's atmosphere is considered so unique that it is one of the things astronomers and exobiologists are looking for in their searches for other earths in distant solar systems.

Photosynthetic bacteria, dating back more than three billion years ago in the fossil record started the process of modifying the composition of the earth's atmosphere. As part of their life process, these bacteria used carbon dioxide and released oxygen as a waste product. The same is still true of modern trees and other green plants. It is a marvelously balanced system, perfectly symbiotic; plants use carbon dioxide from the atmosphere for their growth and release oxygen as a waste product. Animals use oxygen and exhale carbon dioxide. Plants depend on animals and animals depend on plants.

People who remember their basic science classes can skip this paragraph, but for the rest of us, a short review of photosynthesis is useful. The basic chemical reaction is the following:

$$CO_2 + H_2O \longrightarrow CH_2O + O_2$$

According to this reaction, given a suitable energy source such as light from the sun, carbon dioxide reacts with water to produce a basic sugar and free oxygen as a waste product. Notice that as part of the process, carbon is stripped from carbon dioxide and incorporated into the plant's tissue.

This reaction, one of the most important known, tells us the basic requirements necessary for plant growth. The three essential ingredients are: *Sunlight, carbon dioxide and water.* Eliminate any one of these and the plant dies. Everyone realizes that a plant will die if kept in the dark or not watered. Because carbon dioxide is an invisible and odorless gas, it is not commonly thought of as necessary for plant growth. Nevertheless, it is just as essential. Remove most of the carbon dioxide from the air and the plant will die.

What happens when the leaves fall off the trees in autumn or when plants die? Decay begins and the solar energy stored in plant tissue is released as the photosynthetic reaction reverses:

$CH_2O + O_2 \longrightarrow CO_2 + H_2O$

A similar process operates during animal respiration, energy is released along with carbon dioxide and water.[3] This is a natural process that is in constant operation around us. As is true of most chemical reactions, adding heat energy increases the rate of the reaction. Thus, we might expect decay to operate faster in warm climates than cold, as had been confirmed. The process also speeds up when wood is burned. The carbon stored in the wood over a long period of time is rapidly returned to the atmosphere.

Carbon from plants can be stored over much longer intervals of time than represented by the wood in a tree through a series of geologic processes involving increased temperature and pressure that convert buried plant material into peat, than lignite, than soft coal and finally anthracite or hard coal. Related processes convert animal remains, including single-celled organisms, into petroleum, tar, crude oil and natural gas. When these so-called fossil fuels are recovered and burned for their energy, the long-delayed return of stored carbon to the atmosphere finally occurs. The process also occurs naturally at a reduced rate through geologic time when deposits of coal or petroleum become exposed to the oxygen in the atmosphere.

So human activity artificially increases the rate at which carbon dioxide is being added to the atmosphere, which is harmful to human health. In addition to causing warming, isn't this an additional reason carbon dioxide is correctly labeled as a pollutant?

Well, certainly that's the logic, and proponents of this view have no trouble at all in finding a number of persuasive political arguments convincing them it is correct, just as opponents are open to other political arguments. Science, however, is, or at least is supposed to be, based on scientific observation and evidence, not a political agenda, and in a number of ways, the scientific evidence for classifying carbon dioxide as a pollutant has problems. Other books consider political arguments so they will not be examined here. Chapter Nine will consider the relationship between temperature and carbon dioxide. This chapter focuses on what science has learned about carbon dioxide, and greenhouse gases, especially effects that are sometimes cited to bolster the case for considering carbon dioxide a pollutant

It is a well established fact that carbon dioxide is a poisonous gas to animals at high enough concentrations, not because it is harmful itself but because it displaces oxygen. It causes an increased rate of respiration at a concentration above 5% by volume, and more than 30% can be lethal.[4] Such concentrations are far greater than the less than 0.04% of carbon dioxide in today's atmosphere.

How about plants? Increased carbon dioxide is harmful to them too, isn't it?

It's a very different ball park when it comes to plants since they require carbon dioxide as a raw material for survival and growth. There is abundant peer-reviewed literature on this subject which shows the effect of increasing carbon dioxide on most plants is much like adding fertilizer, the rate of growth and productivity is increased. This effect has been observed in such diverse plants as pines, oaks, ginseng, wheat and sweet potato.[5, 6, 7, 8, 9] Other scientists have found that naturally elevated carbon dioxide increases the ability of trees to withstand drought. Even extreme levels of carbon dioxide, as much as a hundred times higher than current, have produced positive responses.

A few peer reviewed studies have shown increasing carbon dioxide produces negative effects in certain plants. For instance, a 1995 article found less stomata development leading to reduced water usage,[10] but improved water use efficiency.[11] Irakli Loladze found plants grow faster with increased carbon dioxide, but with a lower micronutrient content.[12]

Global warming alarmists seem to consider the generally positive effect of increasing carbon dioxide on plant growth and productivity as a climate myth perpetuated by right wing groups. David Chandler and Michael Le Page summarize this view.[14] The article discusses other harmful effects such as ozone depletion and acidification of ocean water and speculates on the future but does not reference a single peer reviewed article to refute the large body of literature reporting boosts in plant growth with increased carbon dioxide. Other articles variously suggest the beneficial effects of increasing atmospheric carbon dioxide may taper off after several years as plants adjust, that the effect is unimportant in tropical areas or that it has negative side effects. In general, however, most peer-reviewed studies continue to show plants respond positively to increasing carbon dioxide.

From a practical standpoint, nursery owners are well acquainted with this relationship. For over a century, they have been adding carbon dioxide to their greenhouses to increase the yields of the various ornamental plants and vegetables they grow.

It might be expected that if this is correct, lowering carbon dioxide would slow plant growth and productivity. This is indeed what a number of studies have found. To cite just one example, rice plants were smaller with less yield when grown in an atmosphere containing approximately half of the current amount of carbon dioxide.

Isn't it true that the amount of carbon dioxide is higher now than anytime in the past 680,000 years? Just that fact alone should be enough to make any sensible person realize something is wrong.

That certainly sounds scary. Of course, that's what climate alarmists hope when they cite such a number. But they always leave out, as Paul Harvey used to say, the rest of the story, a story that a geological perspective on climate change makes clear.

The rest of the story is that the amount of carbon dioxide in the atmosphere today is *abnormally low*, not high, when considered from the viewpoint of earth history and geologic time. To understand how this could be, a little background is required.

The oft-cited 680,000-year figure (or similar number) comes from drilling into ice sheets at first in Greenland, and now in Antarctica, extending the record back to 800,000 years. The technique, once controversial but now generally accepted, involves collecting core samples from the surface deep down into the ice. The ice is not a uniform mass but instead is layered, with layers forming annually, somewhat like tree rings, and each representing a year's worth of snowfall. During the process by which pressure converts snow into ice, tiny air bubbles become entrapped, containing samples of the atmosphere for the year in which the snow fell. These air bubbles can be collected and analyzed giving paleoclimatologists a record of changing temperature and atmospheric composition through several hundred thousand years of time. A more complete explanation of this work as well as how glaciers form is in Chapter Ten.

Chemical analysis of the air bubbles enables paleoclimatologists to determine, for example, the amount of oxygen, carbon dioxide, methane and other gases in the atmosphere many thousands of years ago. Counting the layers down into the ice and back into time gives the age of each layer. At the time this is written (September 2009), a continuous record from the present back to 800,000 years ago is available by combining the results of various studies into a single plot.[16] Carbon dioxide ranges from about 0.0175% (175 parts per million by volume, ppmv) 670,000 years ago to just over 0.03% (300 ppmv) 320,000 years ago. This makes it clear that the current 0.0388% concentration of carbon dioxide in the atmosphere is the highest for the length of time the ice core records represent, several hundred thousand years.

It's obvious that there must be something very unusual about the present, and what's different is humans, the industrial revolution, the burning of fossil fuels. Isn't that right?

The ice core data is cited to make people think that carbon dioxide is abnormally high and that modern industrialization is the reason. Citing the data is effective because knowing no more than the results of ice core drilling, this conclusion seems to make sense. However, a geologic perspective shows us that the current level of carbon dioxide in the atmosphere is actually abnormally low compared to most of geologic time.

For most of the history of the earth, carbon dioxide was present in the atmosphere at much higher concentrations than today and at much higher concentrations than even the most extreme climate alarmists think will occur. During the early part of the Paleozoic Era, from about 550 to 350 million years ago, while life was restricted to the oceans, carbon dioxide fluctuated from ten to twenty-five times higher than today's levels. It dropped sharply in response to the appearance and spread of land plants about 350 million years ago, but shot up again during the Mesozoic Era, the Age of Dinosaurs. From a peak value of more than ten times current levels during the early Mesozoic, carbon dioxide began a long, gradual and irregular decline until the low levels seen in the ice drilling were eventually reached.[17]

Several points should be emphasized regarding past carbon dioxide abundance in the atmosphere of the earth. First, the study cited is based on a geochemical model, Geocarb III, using numerous estimated factors as well as study of past geologic materials, such as ancient soils and sedimentary rocks. It is but one of several such studies. They vary in the factors used, the way the factors are treated and the numbers obtained, but agree on the general trends and conclusions. For most of its existence, it can be stated with assurance that the earth's atmosphere had far more carbon dioxide than today. The spread of green plants in the Paleozoic caused this to drop, but it shot up again during the time of the dinosaurs and then slowly declined over the next two hundred million years until reaching the low level of carbon dioxide in today's atmosphere.

To fit this into the context of the modern global warming theory, it should be clear that if one's purpose is to make people think carbon dioxide is anomalously high today and that humans are to blame, emphasize the ice core data. Keep quiet about how low current carbon dioxide really is compared to the geologic past. By all means, don't say the Mesozoic was an age of luxuriant plant growth, even in Greenland and Antarctica, as the fossil record clearly shows, not an overheated disaster. Far from being a time of crisis, life flourished until an asteroid smashed into earth 65 million years ago. By that time, carbon dioxide had already fallen considerably from its peak.

Another point to emphasize is that a major ice age occurred about 450 million years ago, when carbon dioxide was reaching a peak, far higher than today. Another ice age gripped the earth about 760 million years, and there were several lesser known ice ages. High carbon dioxide in the atmosphere during those times did not prevent these ice ages. There were also times when carbon dioxide was high and the world was what might be described as a tropical paradise, such as the age of dinosaurs, but carbon dioxide was decreasing throughout the Mesozoic. A geological perspective demonstrates with striking clarity the lack of correlation between global climate and atmospheric carbon dioxide.

It is also important to understand that carbon dioxide levels were never steady and unchanging. They have fluctuated throughout the history of the earth. There were no people for most of this time. People didn't cause carbon dioxide to rise and fall, to reach peaks, then decline, only to increase again. Burning coal and oil didn't cause the 200-million-year gradual decline to the low levels of today.

It's hardly a revelation that other sources of carbon dioxide exist besides our overuse of coal and petroleum. The carbon cycle is well known, but the problem is that all our cars, jets, factories, buildings and houses pump so much carbon dioxide into the air that the carbon cycle can no longer maintain a balance. The amounts are huge--seven, eight, maybe ten billion of tons of excess carbon dioxide spewed into the air each year. We do that, not the carbon cycle. Since we caused the problem, we have to fix it, isn't that right?

Each person exhales about 2.3 pounds of carbon dioxide every day and there are about 6.5 billion people. A simple calculation shows that simply by being here, humans contribute roughly 2.5 billion tons each year, a fact global warming alarmists never mention.

The numbers cited are in the correct range for estimates of annual emissions of carbon into the atmosphere according to the U.S. Department of Energy.[18] Calculation shows that carbon dioxide is even higher, more in the range of 32 billion tons per year. Huge numbers certainly, both of them, which is one of the problems in trying to get a grasp on the whole carbon dioxide question. It's even harder trying to understand the carbon cycle, but necessary for understanding the importance of our industrial activities. The following discussion uses carbon instead of carbon dioxide.

Carbon is a scarce element on the earth accounting for only 0.19%, 100 times less than the amount in living things. Life is essentially a process that concentrates carbon in various life forms to use over and over, perhaps the ultimate form of recycling. This is what is meant by the carbon cycle. Much as money can be stored in a vault, carbon is stored in "sinks" in various form on and in the earth. These sinks are reservoirs of carbon, where carbon stays for various periods of time, some short term and some long term. The table below, in GigaTons (billions of tons) of carbon, shows estimates of the major sinks:

Atmosphere	800
Soil Organic Matter	1600
Oceans	40,000
Plants	600
Fossil Fuels	4000
Limestone	100,000,000

Although the numbers are huge, it is immediately apparent that compared to the earth's limestones, the other sinks are insignificant, amounting to only 0.47% of the total estimated amount of carbon. In round figures, limestone and, to a far lesser extent other sedimentary rocks, store almost 99.5% of the total.[19]

Storage of carbon is less important than how carbon dioxide is released into the atmosphere. Humans are the major source for that, aren't they?

That's much too simple a generalization. There are actually many paths for carbon to make its way into the atmosphere after reacting with oxygen to form carbon dioxide. For example, approximately 50 GigaTons of carbon enters the atmosphere from animal and plant respiration and 60 GigaTons comes from decay of organic matter,[20] both much larger than the 8 GigaTons from our use of fossil fuels for energy to power just about everything. Of course, carbon dioxide does not just remain static once it's in the atmosphere. Plants extract it during photosynthesis, and it reacts with the water in streams, lakes and oceans to form carbonic acid, H_2CO_3, which then reacts with various rocks and minerals during chemical weathering, especially limestone. In fact, the various processes that remove carbon dioxide from the atmosphere are so efficient, that the entire amount cycles every 4.5 years.[21]

One aspect of the carbon cycle that is regularly ignored on web sites discussing global warming is what happens during the chemical weathering of rocks. To understand this, the discussion has to become a bit more technical, but the chemistry is held to a bare minimum.

Weathering involves chemical reactions that occur when the minerals composing rocks are exposed to water and the atmosphere at and near the surface of the earth. Especially important in this regard is limestone, one of the most widespread sedimentary rocks, which covers large areas of earth's surface and much of the sea floor in the form of carbonate sediments. Because limestone is composed principally of the mineral calcite, chemically, calcium carbonate or $CaCO_3$, it is highly susceptible to the corrosive action of carbonic acid, produced when carbon dioxide dissolves in water, as it does every time rain falls. Yes, that's right, rain is naturally acidic. Textbooks in beginning-level geology list the following reaction:[22]

$CaCO_3 + H_2CO_3 ---> Ca^{2+} + 2HCO_3^{1-}$

According to this reaction, limestone reacts with carbon dioxide yielding calcium and bicarbonate ions dissolved in water. This is a simplified version of the process, omitting intermediate steps. The complete chemistry of the reaction is available here.[23] The importance of this reaction can be gauged by the fact that it is responsible for forming limestone caves, caverns and sinkholes all over the earth.

What happens to the dissolved ions that result from this reaction depends on the local conditions. The reaction can reverse forming more calcium carbonate. In caves, this forms stalactites and stalagmites. Marine invertebrates also use the reverse reaction to produce their shells. Perhaps even more importantly, hydrated carbon dioxide diffuses out of solution and into the atmosphere. This means that when limestone undergoes chemical weathering, carbon dioxide may be added to the atmosphere. To some extent, this is countered by the use of carbon dioxide from the atmosphere during the weathering of igneous and metamorphic rocks containing calcium and silicon.[24] In terms of area of surface exposure, these rocks are far less important than limestone.

An important basic chemical principle is that most chemical reactions operate faster with increasing temperature. An everyday example of this principle in action is the common experience of having to stir ice tea with vigor to dissolve a sweetener, while only a light stirring suffices in hot tea or coffee. A more extreme example is that wood in a fireplace doesn't oxidize fast enough to produce noticeable heat until one starts a fire. Raising the temperature by a large amount greatly increases the rate of oxidation. We call it combustion or simply burning.

A general rule of thumb is that the rate at which many chemical reactions operate doubles for every 10 degrees C increase in temperature (18 degrees F).[25] It is generally accepted that the earth has warmed by approximately one degree F since the end of the Little Ice Age in the nineteenth century. As a rough approximation, this should have increased the rate of rock weathering by something on the order of 10%. Because limestone accounts for about 99.5% of the stored carbon, even a tiny increase in the rate at which carbon is released from this sink could easily overwhelm the amount we release from burning carbon-based fuels.

Still, though, isn't the burning of fossil fuels the most obvious source of carbon dioxide?

Scientists must be careful to separate fact from speculation. Since 1959 when the Mauna Loa carbon dioxide observatory started in Hawaii, carbon dioxide has increased in the atmosphere from about 0.0316 to about 0.0388% (316 to 388 ppmv). That is a fact. It leads us to another fact, that the carbon cycle is not in balance in terms of the annual addition and removal of carbon dioxide. At the present state of knowledge, we can only speculate as to the cause or causes. The burning of fossil fuels is one possibility, but there are several others. The most obvious or simplest answer may not be the right one.

It must also be emphasized that increasing the carbon dioxide in the atmosphere does not automatically lead to increasing temperatures in spite of media stories assuming the two are inseparably linked. The reasons underlying this statement will be explored in much greater depth in Chapter Nine.

There is a well known example that proves the opposite. In fact, it proves too much carbon dioxide can lead to a runaway greenhouse. It's our sister planet, Venus. The atmosphere is mostly carbon dioxide and the surface is hot enough to melt lead.

The media often mentions Venus with the implication that the same thing can happen on earth if we don't change our evil ways and stop burning coal and gas. They always fail to mention one simple fact, that Venus is almost one third closer to the sun than we are. That by itself is enough to raise the surface temperature on Venus almost to the boiling point of water.[26]

This fact has tremendous implications for us. It means that water most likely never condensed on Venus to form oceans. Even if oceans once existed, they could not have lasted long before being evaporated into the atmosphere. There the intense ultraviolet radiation would have broken the water vapor down into hydrogen and oxygen, and it would have been lost into space. On earth, the geologic record shows that oceans formed very early and have continually existed ever since. They absorb a major portion of the carbon dioxide that's released into the atmosphere each year, but as discussed earlier in this chapter, the really important sink for storing carbon is limestone, which forms only in water. Without liquid water, there would be no limestone to store all the excess carbon locked up as $CaCO_3$.

Venus is in many ways our twin--almost the same size and density for example. It also has just about the same amount of carbon as the earth, only here it is stored safely in limestone, but on Venus it's stored in the atmosphere, accounting for 96.5% of the atmospheric gases. The atmosphere is so thick that the surface pressure is enough to crush a submarine, 91 times greater than surface pressure on the earth.

In spite of what global warming alarmists imply, this cannot happen here as long as the earth remains the same distance from the sun. This is confirmed by the much higher concentrations of carbon dioxide in the atmosphere of the earth at various times in the past, discussed earlier in this chapter, yet there was no runaway greenhouse effect. It is true that the earth was much warmer than today during some of the times when there was more carbon dioxide in the atmosphere, but at other times, it was much colder. It is difficult to spot any correlation between the two.

SUMMARY

Although the EPA considers carbon dioxide to be a pollutant, it is actually a natural waste product of animal respiration. Each time we breathe in oxygen and then exhale, we are releasing carbon dioxide, about 2.3 pounds for each person per day. In a marvelously balanced symbiotic system, plants take carbon dioxide from the air, combine it with water and, given a suitable energy source such as sunlight, produce basic sugars by photosynthesis, releasing oxygen as a waste product into the atmosphere.

Contrary to popular opinion, for most of the time the earth has existed, its atmosphere contained far more carbon dioxide than it does today. Scientists think carbon dioxide was an important gas in the chemically reducing atmosphere that characterized the early earth. Although it also contained nitrogen, as the modern atmosphere does, it did not have enough oxygen to support respiration. Oxygen slowly built up after photosynthesizing bacteria appeared 3.2 billion years ago or even earlier.

When plant and animal tissue undergoes decay, carbon dioxide is returned to the atmosphere. Given required geologic conditions, including burial to produce a certain amount of heat and pressure, plants can be converted into coal and animals into forms of petroleum, storing carbon. When exposed near the surface, the carbon in these materials is oxidized, releasing carbon dioxide. Recovering such deposits and burning them for their energy accelerates the process.

Carbon dioxide is an invisible, odorless gas that is lethal to animals in amounts high enough to displace oxygen. Contrary to the way it's usually portrayed, higher concentrations of carbon dioxide in the atmosphere are good for plant growth. The scientific literature supporting this conclusion is voluminous, generally finding increased rates of growth and increased productivity in diverse plant groups, including vegetables, grains and trees.

On the earth, the carbon cycle circulates and stores carbon in various forms. Its basic understanding is essential for understanding climate change. Respiration and photosynthesis are important parts of this cycle as is long-term storage of carbon in "sinks," such as dissolved in the oceans and in soil organic matter. Although inexplicably often omitted from most diagrams of the carbon cycle, limestone rock is by far the most important sink, accounting for about 99.5% of the total stored carbon.

The stupendous quantities of carbon stored in sinks are hard to comprehend. The huge numbers must, however, be considered if one is to put into proper perspective the approximate 8 billion tons (GigaTons) of carbon added to the atmosphere each year from using coal and petroleum. Climate alarmists bandy about such numbers for effect omitting that they represent only tiny fractions of the carbon entering the atmosphere from natural sources, only 16% of the amount from animal respiration for example. Any process changing the rate at which carbon is added to or removed from the various sinks could easily render insignificant the amount from fossil fuels. To consider one possibility, a slightly warmer climate might add more carbon dioxide from the chemical weathering of limestone than our use of coal and oil.

The elite media has sensationalized the fact that in the last 150 years or so, carbon dioxide in the atmosphere has been increasing, and that it is now higher than any time in the past several hundred thousand years. This is supposed to sound damning, as indeed it does, unless one realizes that for most of geologic time, our atmosphere contained far higher levels of carbon dioxide than today. In fact, from a geologic perspective, today's atmosphere is abnormally low in carbon dioxide, not high. From a peak value of over 10x more carbon dioxide in the atmosphere 200 million years ago when dinosaurs dominated the earth, the concentration has fallen to the extremely low amount that characterizes the last several hundred thousand years. It goes without saying that humans had nothing to do with this huge variation. It is also true that little or no correlation can be noted between the temperature in past geologic periods and the estimated amount of carbon dioxide in the atmosphere.

It verges on propaganda to say, as climate alarmists do, that if we don't become better stewards of the earth, we might produce a runaway greenhouse effect that would make our world as uninhabitable as Venus. The fact that far higher amounts of carbon dioxide in our atmosphere were the norm for most of geologic time without this happening proves the fiction of such a scenario. The hellish conditions of Venus resulted directly from that planet being one third closer to the sun than we are, causing the surface temperature to shoot up to near boiling, evaporating all the liquid water. Without oceans, no limestone formed, leaving the atmosphere as the major sink of carbon dioxide. Move the earth this close to the sun and the process would happen here, but not because we power our jets, cars and power plants with fossil fuels.

NOTES AND SOURCES

(1)Available: http://www.time.com/time/health/article/0,8599,1887263,00.html
(2) Much useful basic data about the earth is available here: http://www.nineplanets.org/
(3) Class notes, University of Washington, Department of Atmospheric Sciences, for the course "Climate and Climate Change," available: http://www.atmos.washington.edu/2002Q4/211/notes_carboncycle.html
(4) Some of the effects of carbon dioxide poisoning are described at this site:http://volcanoes.usgs.gov/hazards/gas/index.php
(5)*Environmental and Experimental Botany*, Vol 34, p. 337, 1994, R.L. Garcia, et al, "Net photosynthesis as a function of carbon dioxide in pine tree growth at ambient and elevated CO_2."
(6) *Environmental Pollution*, Vol. 147, p. 516, 1994, E. Paoletti, et al, "Photosynthetic responses to elevate CO_2 and O_3 in *Quercus ilex* Leaves at a natural CO_2 spring."
(7) *Plant Physiology and Biochemistry*, Vol 43, p. 449, 2005, M.B. Ali, et al, "CO_2-induced total phenolics in suspension cultures of *Panax ginseng* C.A. Mayer roots: role of antioxidants and enzymes"
(8) *Advances in Space Research*, Vol 42, p. 1917, 2008, L.H. Levine, et al, "Physiologic and metabolic responses of wheat seedlings to elevated and super-elevated carbon dioxide."
(9) *Journal of Plant Biotechnology*, Vol 7, p. 67, 2005, J.A. Teixeira de Silva, et al, 2005, "Microprogation of sweet potato (*Ipomoea batatas*) in a novel CO_2-enriched vessel."

(10) *New Phytologist*, Vol 131, p. 311, 1995, F. Woodward and C. Kelly, "The influence of CO_2 concentration on stomatal density."

(11) *Annual Review of Plant Molecular Biology*, Vol 48, p. 609, 1997, B.G. Drake, et al, "More efficient plants: a consequence of rising atmospheric CO_2?"

(12) *Trends in Ecology & Evolution*, Vol. 17, p. 457, 2002, Irakli Loladze, "Rising atmospheric CO_2 and human nutrition: toward globally imbalanced plant stoichiometry?"

(13) *Plant Cell and Environment*, Vol. 21, p. 613, 1998, R. Tognetti et al, "Transpiration and stomatal of *Quercs ilex* plants during the summer in a Mediterranean carbon dioxide spring."

(14) *New Scientist*, p. 17, May 16, 2007, David Chandler and Michael Le Page, "Climate myths: higher CO_2 levels will boost plant growth and food production."

(15) *Journal of Plant Physiology*, Vol. 157, p. 235, 1995, R.W. Gesch et al, "Subambient growth CO_2 leads to increased Rubisco small subunit gene expression in developing rice leaves."

(16) The combined plot is available here: http://en.wikipedia.org/wiki/File:Co2-temperature-plot.svg

(17) *American Journal of Science*, Vol. 301, p. 182, 2002, Robert A. Berner and Zavareth Kothavala, "Geocarb III: A revised model of atmospheric CO_2 over Phanerozoic Time."

(18) The Carbon Dioxide Information Analysis Center is the branch of the U.S. Department of Energy responsible for gathering and publishing data on national and world-wide emissions of carbon and carbon dioxide. See the following link: http://cdiac.ornl.gov/trends/emis/meth_reg.html

(19) Estimated quantities of carbon stored in various sinks are shown here: http://www.eoearth.org/article/Carbon_cycle

(20) *ibid*

(21) *The Dynamic earth*, Second Edition, John Wiley and Sons, Inc., New York, p. 512, 1992, Brian J. Skinner and Stephen C. Porter.

(22) *ibid*

(23) For the reaction between calcium carbonate and hydrogen ions in water, see: http://antoine.frostburg.edu/chem/senese/101/inorganic/faq/limestone-and-water-reaction.shtml

(24) *Greenhouse Gas Sinks*, Publisher CAB International, p. 86, 2007. Dave Reay et al.

(25) See http://www.geographypages.co.uk/weathering.htm.

(26) *The Nine Planets* is an excellent web site full of useful and reliable information about the solar system. Their presentation on Venus is here: http://www.nineplanets.org/venus.html

CHAPTER THREE—NORMAL CLIMATE AND CLIMATE CHANGE

Of course, climates have changed in the past, but as a geologist, wouldn't you agree that the speed with which the climate is changing now is unusually rapid, even unprecedented?

Since this book takes a geological perspective on climate change, the subject of change in general needs to be addressed before delving into climate.

There are worlds where not much of anything ever happens. We don't happen to live on one of those worlds. Geologically speaking, our lovely blue planet marbled with snow white clouds, is an unusually active world. It is alive with earthquakes, colliding plates, volcanoes, rifting continents, spreading oceans, tsunamis, floods, heat waves, blizzards, droughts, rock slides, avalanches and yes, changing climates. For a geologist, it would be hard to find a more interesting planet.

We don't have to look very far to find a world where this is not true. When one looks up at the full moon and sees the dark and light splotches, what used to be called the man on the moon, it's not apparent that it is a dead world. Looking through binoculars and seeing an apparently endless expanse of craters, it seems fascinating and alive. We now realize all these craters are due to rocks from space smashing into the surface, small and gigantic, but they were once thought to result from volcanic activity. No volcanic activity of any type has ever been spotted on the moon. The fascinating surface we look at is little changed from the time it formed over 3.5 billion years ago. Nothing happens there. It's a world eternal, frozen in time, static, boring, waiting long intervals of time for the next random strike of some wayward meteor.

The Apollo astronauts left seismic detectors on the moon to record any "moonquakes" that might occur after the missions ended. As it turned out, for the several years the detectors operated, they detected each year about 3000 moonquakes from several hundred miles down. These, however, are far too gentle to have been felt at the surface. Their total energy is less than one ten-billionth of the amount released in earthquakes during the same period.[1] The astronauts' footprints are still as fresh and pristine as the day they climbed into their space capsule and headed back to a much more interesting world.

The term "normal climate" is used correctly in reference to the moon. It has a climate unimaginably harsh. Temperature on the moon's surface varies over the course of a few weeks from far colder to far hotter than the earth has ever experienced during its entire 4.56 billion-year history. That is the normal climate for our close neighbor in space and it does not seem to change much.

In contrast, normal climate is meaningless for the earth in a geologic sense because its climate is always changing. Climate changed in the past, it is changing now and it will change in the future. It changed before humans were here and it will change after we are gone. Change in climate is constant, but not just climate, all sorts of geologic change. As geologists learn early in their education, the only thing constant on the earth is change itself.

Of course, variations in how fast climate changes have occurred. There have been times in earth history when the planet seemed to be endlessly locked into a pattern of static climate, for example, 50 million years of a frozen planet about 760 million years ago, the time geologists sometimes jokingly call "snowball earth." Eventually, however, this passed and the climate warmed.

The question, however, concerned the most recent episode of warming starting in the mid to late nineteenth century following several hundred years of cooler conditions during what geologists often refer to as the Little Ice Age, and specifically, whether this recent warming is unprecedented.

The answer is that it is not unprecedented at all. Viewing the twentieth century warming from a perspective of a few thousand years clearly shows it to be a rather minor fluctuation during the current interglacial epoch, an Eden-like time of generally warm climate.

Maybe this is the view of a geologist, but it is not supported by recent science. Consider that the June 2009 issue of *Discover* published "The Big Heat," an article interviewing four scientists concerning global warming. One of them stated that we are seeing rates of warming, "that exceed natural rates by a factor of 100." Isn't this correct?

The article in the magazine is typical of how most of the national media treats global warming. Not one single reference to peer-reviewed articles published in scientific journals is given anywhere in the piece. In contrast, there are numerous such articles that show the statement is wrong. Consider the Antarctic ice core data.[2] It shows several episodes of warming at rates far exceeding what has occurred during the past 130 years or so. In fact, this ice record shows that the warmest part of the current interglacial was several thousand years ago, not today. A high resolution graph of the Vostok temperature record (Figure 1) shows this clearly.[3] The same conclusion is reached from studies which look at hundreds of years of temperature rather than thousands. One such study is for the northeastern and north central United States.[4] The general

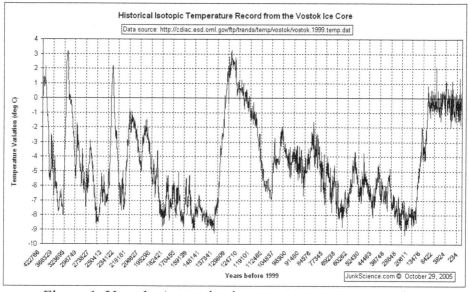

Figure 1. Vostok, Antarctica ice core temperature record.

pattern found is warmer temperatures several hundred years ago before the Little Ice Age and no abrupt warming in the past 130 years. To cite just one more study from the large number that have been published, consider the 2,000-year record of temperature for the northern hemisphere, derived using a combination of methods, from NOAA's Paleoclimatology Program (Figure 2).[5] The temperature plot[6] shows a several-hundred-year general rise in temperature beginning around AD500 and reaching a peak shortly after 1100, when the climate began to cool as it slid into the Little Ice Age. The climate started to warm again in the nineteenth century. The study found the current warming similar in the scale and rate to that which produced the well known warmth that peaked around AD1000-1100 that climatologists refer to as the Medieval Warm Period. Numerous studies suggest temperatures were about a degree F warmer than today, warm enough for Vikings to establish colonies in Greenland, where crops could be grown along the southern coast.

Using ice core data to look at the previous 5000 years would produce an even clearer picture. This would show that temperatures warmer than today were reached more than once during that time. Regarding the magazine article, unless a peer reviewed study can be cited showing temperature increasing 100 times faster than natural rates, with sound

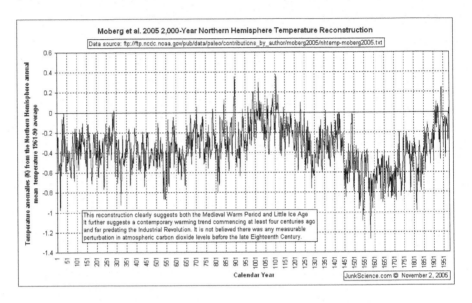

Figure 2. NOAA 2000-year northern hemisphere temperature reconstruction.

science used to establish natural rates, the statement would seem to be more hyperbole than science.

You referred to "natural rates" and earlier, you mentioned "normal climate" saying the term doesn't apply to the earth. This term is often encountered in climate literature, but what exactly does it mean?

As a geologist, the term seems odd because how does one go about deciding what is normal climate? Is the climate of the Little Ice Age any less normal than the climate of the twentieth century? How about the cold of the Pleistocene Ice Age, often referred to as the great ice age, or the lack of any ice caps anywhere in the world when dinosaurs lived? Are these normal or abnormal? How is one to choose, especially since the climate is always changing?

The problem, however, is not one's choice of viewpoint, but confusing the term "normal climate" with a term widely used in climatology, "climatic normal." Climatologists invented and have been using the concept of a climatic normal for 150 years. It is simply the numerical average of some climatic element, such as temperature, over an arbitrary interval of time. A thirty-year period of time has been used since 1933, based on agreement reached at the International Meteorological Conference in Warsaw. Thirty years was deemed long enough to smooth out short-term fluctuations, but short enough to spot trends.

As used by the U.S. National Climatic Data Center (NCDC), climatic normals are updated every ten years through the end of the previous decade ending in zero, such as the currently used 1971-2000 interval.[7] When more recent data becomes available, this will shift to 1981-2010 interval. The World Meteorological Organization uses what are called *standard normals* computed every thirty years, rather than every ten, 1931-1960, 1961-1990, etc.

It is important to understand how climatologists use climatic normals to avoid being misled. This has been a problem long before global warming became an issue and was discussed by the Director of Climatology in 1955: "The layman is often misled by the word. In his every-day language the word normal means something ordinary or frequent....When (the meteorologist) talks about 'normal', it has nothing to do with a common event.... For the meteorologist the word 'normal' is simply a point of departure or index which is convenient for keeping track of weather statistics.... We never expect to experience 'normal' weather."[8] It is erroneous to expect the "normal" temperature for a certain month to be the temperature one experiences during that month. As discussed previously, wide swings and extremes can hide in average values to the point of becoming invisible.

The NCDC advises that it is a mistake to draw conclusions by comparing various 30-year periods because of changes in computation, instrumentation and how the data was obtained. Global warming alarmists routinely ignore this caution, treating changes from 30-year norms as reliable proof of long term climate trends despite the NCDC's admonition that, "The differences between normals due to these non-climatic changes may be larger than the differences due to a true change in climate."

Charts showing global warming are typically constructed using a temperature base period of 1951- 1980. Why is that time considered normal climate instead of the more recent 1971-2000?

It is correct that GISS (Goddard Institute for Space Studies, a division of NASA) as well as other agencies that monitor global temperature, commonly use this 30-year period as their reference. Graphs are typically constructed showing temperature departures from the mean of this period, drawn as a straight line at zero. Studying a recent GISS version (Figure 3),[10] four distinct regimes of temperature are apparent: 1) cool temperatures from the beginning of the chart in 1880 to about 1910; 2) warming from 1910 to the mid 1940s; 3) thirty years of cooling from the mid 1940s to about 1976; 4) warming beginning in the late 1970s.

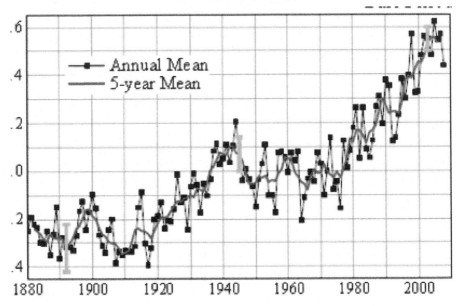

Figure 3. GISS global temperature, 1880-2008.

It is hotly debated whether the most recent warming trend still continues or whether it peaked a few years ago, ushering in a cooling trend. This question need not concern us now, but will be revisited in more detail in a later chapter.

The use of this particular 30-year norm might be an historic artifact from the 1990s when global warming caught on as a cause and began to become widely known. At that time, it was the most recent data available, but this explanation does not account for continuing to use this 30-year period after more recent data became available.

Another possibility, widely believed by global warming skeptics, stems from the fact that the period 1951-1980 was a period of generally cooler climate. Taking this period as "normal climate" greatly emphasizes the subsequent warming, making it easier to convince undecided minds. Whether this is the intention or not is another hotly debated topic, but wide media coverage referring to this 30-year period as "normal climate" certainly helps to steer the public's collective mind toward accepting that today's climate is abnormal.

A chart constructed using a 1971-2000 base would still show the same amount of warming from 1880, wouldn't it?

It would, but it wouldn't stick up as high above the zero line, so, on first glance, it would appear to be less, and first impressions are very important in winning public support.

Aside from choice of base line, from a geological standpoint, beginning a global temperature chart in 1880 comes with a built-in bias exactly in the direction global warming alarmists want in order to convince the public that a crisis exists. Doing so is very convenient for their argument, because it allows them to say that warming started right at the same time as the industrial revolution was coming on strong. Reading Charles Dickens shows that coal smoke and soot were all over England, and it was much the same in the northeastern U.S. They argue that for the first time in the history of the earth, humans began to influence the atmosphere in a big way by dumping large amounts of carbon dioxide into it, causing warming.

Starting the record in 1880 cuts out hundreds of studies plus widespread historical records from Europe and the United States showing several centuries of colder climate before 1880, often called the Little Ice Age. It came to an end in the late nineteenth century but, as mentioned earlier, was proceeded by the medieval warming, several centuries with climate as warm or warmer than today. Why not show the full millennium? That would show clearly that climate change occurred before people had the ability to muck it up. The difference is clear from looking at the two previous illustrations. The public would see a thousand years of climate change as a really long-term record, although, to a geologist, it is still very much short term. Making it better known would at least give a clearer picture of climate change than starting only at the tail end in 1880.

Instrument records only go back to 1880. Isn't that the reason for starting at that time?

Thermometer records become more abundant around that time, but some go back to the eighteenth century. There are good thermometer records for Europe back to about 1750, with some even earlier. While it's true some of these early records have problems, the same is true of newer records.

Just because instrumental records don't exist for the Little Ice Age or the Medieval Warm Period does not mean we are in the dark about what the climate was like at those times. Geologists, geochemists and paleoclimatologists have developed a number of indirect techniques for learning about past climates and climate change. Known as proxy temperatures, most of these methods are not direct measures of temperature, but they can inform us about past climates. Some of them are best suited to relative short time scales of perhaps a few hundred years, others to thousands of years and some to millions and even hundreds of millions of years. Caution must be used in applying them to determine whether some other factor than climate might have been causing a variation in the factor being measured. An assumption inherent in using proxy temperature measures is that a particular proxy's response to varying environmental conditions observed today also operated in the same way during past geologic periods. This is a basic principle used throughout the geological sciences and is soundly based on the constancy of physical and chemical laws in space and time, a basic tenant of all science.

As an example of a proxy, if pollen for a warm weather plant is recovered from a deep layer of sediment in a lake, but we see less of that particular pollen in the younger layers above it and more of a cold-adapted species, we have strong evidence that the climate was becoming colder as that particular sequence of sediments accumulated at the bottom of the lake. Studying the temperature regimes that similar species live in today allows making good estimates of the water temperature. Climate information can be obtained from oceanic sediments by noting the abundance of cold and warm water-adapted fossils of tiny shelled invertebrates known as foraminifera. Plant and animal fossils provide similar data on land. Consider the implications for climate if we find palm tree fossils in northern Greenland in strata a hundred million years old or wooly mammoth in New Jersey from 25,000 years ago.

Tree rings are probably the most generally known proxy temperature method. Dendroclimatology, the study of the relationship between past tree ring growth and climate, is a widely used technique to learn about past climate changes. Two obvious factors control tree ring width and density: air temperature and rainfall; however, a number of other factors can play a role. These include soil temperature and moisture, the number of sunny days, the amount of wind, competition from other plants, insect attacks and tree age.

By carefully measuring the ring pattern of many trees in an area, dendroclimatologists can establish a chronology that reaches back to the age of the oldest tree growing in that area. Scientists can extend the record further back by matching the ring pattern found in older wood, such as can be found preserved in buildings, flood plains, peat bogs etc. One looks for a particular portion of the ring pattern (ring width and density) that matches part of the pattern from living trees. This is referred to as *correlation*. The tree ring record has been extended back many thousands of years using this methodology.

Many other types of climate proxies are being used. Ice cores from drilling into ice caps in Antarctica and Greenland have already been mentioned and will be discussed in greater detail in Chapter Nine. Good introductions to using proxy data in paleoclimatology are available on the web including JunkScience.com.

Climate skeptics argue that global warming is nothing more than a natural fluctuation in climate and that temperature proxy data support this contention. Is this correct?

Geology started making use of what we now call proxy climate data in the nineteenth century well before other sciences. Early on, the work showed that climate not only had changed in the past, but the changes were immense. Many places in Europe had huge boulders, seemingly dumped at random, and giant piles of gravel scattered about. The prevailing wisdom was that such deposits were evidence of the great Biblical flood of Noah, but the work of Louis Agassiz showed that these features as well as many others were characteristic of glaciers that existed in the high valleys and peaks of the Alps. He argued convincingly that in the past, large portions of Europe had been covered by the same sort of glaciers. After he came to the United States in 1846, he was able to point to the same sort of features as evidence that the ice had also covered the northern portion of the U.S.

What eventually came to be called the Pleistocene Ice Age, for the most recent Epoch in earth History in which it occurred, at first was thought to be a single cold epoch of climate change. Additional field study, however, began to yield clues that this view was mistaken. During the first half of the twentieth century, four advances of the ice were accepted, but climate proxies developed in the second half of the century showed that this still seriously underestimated the number of times the ice had advanced south across the continents. Geologists now recognize that over the past several hundred thousand years the ice advanced and retreated in a remarkably regular pattern roughly every 100,000 years. In other words, climate change during the geologically recent past had been regularly cyclic.

Other climate cycles are now known, both shorter and longer than the glacial pattern in the Pleistocene. Perhaps best known is the eleven-year sunspot cycle which has been observed since Galileo's time. It is now well established that total solar irradiance (TSI) from the rough average of 1,366.5 watts per square meter (W/m^2) varies about 0.1% during this cycle, but it is debated whether this is enough to affect earth's climate.

TSI is greatest during the part of the cycle when sunspots are at a maximum because the dark splotches are accompanied by *faculae*, bright areas emitting more energy. Both sunspots and faculae increase during times of greater solar activity due to an increase in the number of magnetic storms on the sun. Increased solar activity is thought to be related to warmer temperatures on earth.

A solar variation of longer duration involving the number of sunspots seems to have a greater effect on climate. In particular, the time from AD1650 to 1715 is notable for its very low number of sunspots. Known as the Maunder Minimum, some climatologists have linked it to the coldest part of the Little Ice Age.[12] Several other sunspot minimums, such as the Dalton Minimum (1790-1820) and the Sporer Minimum (1450-1550) have also been linked to colder conditions, but none were as deep as the Maunder. Solar irradiance, high during the last several sunspot cycles, became low in 2006 and has remained at the lowest level since satellite data became available in the early 1970s. Already more than two and a half years past due as this is being written (September 2009), some astronomers raise concerns that a new sunspot minimum might be starting.[13] The lack of any global warming over the past several years supports this.

Kyle Swanson of the University of Wisconsin-Milwaukee thinks the slight cooling the globe as a whole has experienced in the last few years suggests a thirty-year cooling trend might be starting. The Discovery Channel quotes him saying, "This is nothing like anything we've seen since 1950. Cooling events since then had firm causes, like eruptions or large-magnitude La Ninas. This current cooling doesn't have one." The amount of cooling, about 0.4 degree F to this point, while not large, stands in sharp contrast to a 30-year warming trend before this. Climate alarmists argue that it is only a minor squiggle on a curve of generally rising temperature. The next few years should make it clear who is right. Many peer reviewed articles have been published concerning other possible climate cycles, ranging in length from decades to millennial scale and even longer, far too many to discuss here. If we examine Figure 4, a chart showing the generally accepted record of climate over the past 550 million years of earth history, the Phanerozoic,[15, 16] it is

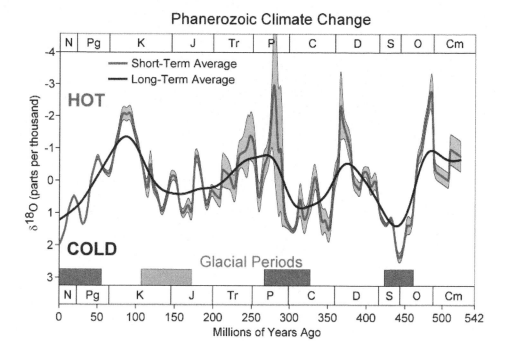

Figure 4. Climate change over the last 550 million years.

apparent that the climate of the earth has oscillated from hot to cold over immense spans of time.

As mentioned previously, many times in the past had climates far warmer than today, with the last great peak being reached in the late Cretaceous Period (K) about 75-80 million years ago. A cooling trend started then, but by the end of the Cretaceous, the earth still had not developed polar ice caps.

The last 65 million years since the end of the Cretaceous and the extinction of the dinosaurs is especially interesting for putting present global climates in perspective. Studying an expansion of the Veizer graph[17] (Figure 5) shows that the cooling of the late Cretaceous was followed by 15 million years of abrupt warming to produce the Eocene Optimum, but cooling resumed after that. The sharpest drop in average global temperature during the entire time was between 35-34 million years ago. Many paleoclimatologists think the Antarctica ice cap formed during this time.

Another long warming trend started after that, reaching a peak about

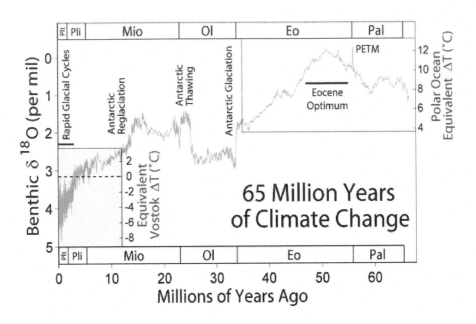

Figure 5. 65 million years of climate change.

26 million years ago. The ice cap had melted back considerably by this time, but it expanded anew during the severe cooling that started about 14 million years ago. Cooling has dominated the last 14-million years of climate history. As the graph shows, the line toward cold conditions is not smooth. It is important to realize that because of the time scale used to construct this graph, many of these "squiggles" represent long intervals of global warming lasting thousands of years. The north polar ice cap formed during this time and the climate had became cold enough two million years ago with the beginning of the Pleistocene that the recent pattern of cyclic glaciation started. During the colder phases, immense glaciers advanced from polar regions into mid-latitudes, but between advances of the ice, warmth returned during the shorter interglacials.

From a geologic viewpoint, the last 65 million years of climate history makes it clear that what is unusual about recent climates is that they have been cyclic and abnormally chilly, rather than unusually warm.

While interesting and certainly adding perspective to recent climate change, modern societies are much more affected by the "squiggles" of climate change than long term trends involving thousands or millions of years. Just because cyclic climate change

occurs does not exonerate modern civilization from causing the recent warming. Isn't that correct?

Without doubt, and by the same token, the convenient end of the Little Ice Age at the time the industrial revolution was shifting into high gear does not mean that natural climate cycles suddenly ended. Oh, sure, climate alarmists admit that *some* of the recent warming could be due to natural causes, but go right ahead talking about global warming as if it's *all* caused by carbon dioxide and other greenhouse gases. It is a fact that climate cycles have operated in the past at all time scales, and it is highly likely that they continue to operate. Another fact is that scientists have a poor understanding of the factors that drive the various cycles, but several different causes are likely to be involved.

A number of recent studies suggest the solar irradiance, the so-called solar constant, may be more variable than once thought and might be important in short term climate variations. Large volcanic eruptions also affect climate over short time scales. Regular variations in the earth's orbit around the sun over tens of thousands of years, the Milankovitch cycles, have been linked to the cyclic pattern of glaciation during the Pleistocene. Global plate tectonics may be important in producing long scale climate change. According to the supercontinent theory, plate tectonics operates in a cyclic fashion over several hundred million years to change the configuration of the earth from one or two huge landmasses (supercontinents) to several smaller landmasses with gigantic oceans.[18] During the part of the cycle when continental collisions are numerous, leading to the supercontinent phase, storage of carbon in limestone dominates and a cooler climate results. Conversely, when volcanic activity greatly increases during the sea floor spreading and ocean formation portion of the cycle, carbon is released from storage in limestone, and the global climate warms.[19]

Of course, these cycles are not affected at all by human activities, so they continue to operate as they did in the past. Huge changes in climate characterize earth history without human input, far exceeding anything in the global warming theory. The possibility that such cycles might still be influencing climate cannot be ignored.

SUMMARY

From a geological perspective, the earth is an extremely active planet. This is in sharp contrast to our close neighbor, the moon, which is geologically dead because it has lost nearly all of its internal heat. Being several times the size of the moon, the earth has retained sufficient interior heat to drive an active plate tectonic system that constantly reshapes the lithosphere, the solid outer shell on which we live. Combined with solar energy that drives a very active weather and climate system, change is the one constant on the earth.

A geological viewpoint is useful for helping to put modern climate change in perspective. Studying earth history shows us that climate changed in the past, it is changing now and it will change in the future. This is not to say that the rate of change remains constant. In fact, there have been long intervals of time in the past when the climate was remarkably stable undergoing only minor fluctuations and other times, when rapid change occurred. As examples, the climate of the Mesozoic Era, when dinosaurs roamed the earth, was generally much warmer than today over millions of years, while rapid change, both global cooling and warming, characterizes the recent Ice Ages of the Pleistocene.

Global warming alarmists argue that modern global warming cannot be due to normal climate change processes because the warming is far more rapid than anything in the past. If true, this would indeed be cause for alarm, but ice core drilling shows the warming since 1880 is not unprecedented at all, either in temperature or the rate of change. Both were exceeded in previous interglacials.

The term "normal climate" is a useful concept in climatology referring to the most recent thirty-year period of data, such as the currently used 1971-2000 interval. Such "standard normals" are updated every ten years with the availability of the next decade's data. Climate alarmists, however, stick to the interval from 1951-1980, a period of cooling. Global warming skeptics charge that plotting temperature change from that time helps to accentuate the warming. As a term, "normal climate" is often misused in a manner that makes the public think today's climate is not normal. Widely publicizing a graph showing global temperatures starting to warm in the late nineteenth century and ascribing it to human uses of fossil fuels without mentioning that a several century long interval of abnormally cold weather known as the Little Ice Age had just ended, is even more egregious.

Thermometer temperature records are scarce from the eighteenth century and almost non-existent before that, but "proxy" data provide climatic information from much earlier times, even back hundreds of millions of years. Among the methods that have been used to obtain such data, pollen in sediment layers, fossils, tree growth rings (dendrochronology) and air bubbles from annual layers in glacial ice have become especially well known, but numerous others have been developed. It is by using such techniques that paleoclimatologists have learned that climate change in the past has often been cyclic, alternating through warm and cold conditions during various intervals of time.

Climatic cycles as short as eleven years, related to the sunspot cycle, and as long as the several-hundred-million-year supercontinent cycle of plate tectonics have been recognized. Longer periods of minimal solar activity, such as the Maunder Minimum from AD1650 to 1715, have been accompanied by decades-long intervals of high solar activity, but data is insufficient to say whether such variations in solar activity is regularly cyclical. Times of decreased solar activity have been associated with colder climate, and warmer conditions with a more active sun (many sunspots). Global warming skeptics point out that the sun was very active during most of the Twentieth Century, but the delay of the next sunspot cycle by more than 2.5 years may be a sign of changing solar activity toward a less active sun.

A particularly significant discovery resulting from proxy data was the cyclic nature of advancing and retreating ice sheets during the Pleistocene Ice Age starting about 2 million years ago. Long time intervals of extreme cold with glaciers advancing from polar regions into formerly temperate zones were followed with remarkable regularity by shorter "interglacials" characterized by a return to warm conditions. The current interglacial, with warm climatic conditions similar to earlier interglacials, started about 18,000 years ago.

NOTES AND SOURCES

(1) *Observing the Moon: the modern astronomer's guide*, Second Edition, p.137, 2007, Gerald North, Cambridge University Press

(2) *Nature*, Vol. 399, p. 429, 1999, J.R. Petit et al, "Climate and atmospheric history of the past 420,000 years from the Vostok ice core, Antarctica."

(3) For the Vostok temperature record, see:
http://www.junkscience.com/MSU_Temps/Vostok_long.html

(4) *World Data Center-A for Paleoclimatology Data Contribution Series #93-001*, NOAA/NGCD Paleoclimatology Program, Boulder, CO, USA, 1993, K. Gajewski, "Little ice age summer temperatures and annual precipitation for northeastern and northcentral United States."

(5) IGBP PAGES/World Date Center for Paleoclimatology Data Contribution Series #2005-019, NOAA/NGCD Paleoclimatology Program, Boulder, CO, USA, 2005, A. Moberg, "2,000-Year Northern Hemisphere Temperature Reconstruction."

(6)Temperature plot for the Moberg study is here: http://www.junkscience.com/MSU_Temps/Moberg2005.html

(7) Climatic normals are discussed on NOAA 's web site: http://lwf.ncdc.noaa.gov/oa/climate/normals/usnormals.html

(8) This quotation is from Dr. Helmuth E. Landsberg, available on the NCDC's web site under "Frequently Asked Questions": http://lwf.ncdc.noaa.gov/oa/climate/normals/usnormals.htmlFAQ

(9) *Ibid*

(10) GISS's most recent chart of global temperature departure from the 1951-1980 norm is available here: http://www.columbia.edu/~jeh1/mailings/2009/20090113Temperature.pdf

(11)NOAA's introduction to proxy data is here: http://www.ncdc.noaa.gov/paleo/primer_proxy.html
Another good discussion is here: http://www.physicalgeography.net/fundamentals/7x.html

(12) NASA has a good discussion of various solar cycles and variations in total solar irradiance available here: http://earthobservatory.nasa.gov/Features/SORCE/

(13) A video of the BBC coverage on the current low solar activity is available here: http://commons.wikimedia.org/wiki/File:Phanerozoic_Climate_Change.png

(14) Dr. Swanson's thoughts on a new 30-year cooling trend is available here: http://dsc.discovery.com/news/2009/03/02/global-warming-pause.html

(15) *Chemical Geology*, Vol. 161, p. 59, 1999, J. Veizer et al, "Evidence for decoupling of atmospheric CO^2 and global climate during the Phanerozoic."

(16) The composite graph from the Veizer article showing the ever changing climate of the Phanerozoic is available at several web sites including: http://en.wikipedia.org/wiki/File:Phanerozoic_Climate_Change.png.

(17) The expansion of Veizer's diagram emphasizing the last 65 million years is available here: http://commons.wikimedia.org/wiki/File:65_Myr_ClimateChange.png

(18)A recommended site for the supercontinent cycle is here: http://maps.unomaha.edu/Maher/plate/week12/super.html

(19) A simple discussion of how plate tectonics might relates global climate is available at this site: http://dilu.bol.ucla.edu/

CHAPTER FOUR—EXAMPLES OF CLIMATE CHANGE

No thinking person denies that climate changed in the past. At times, the earth got warmer, but it got warm over eons of time. There was plenty of time for plants and animals to adjust. That's not true anymore. Is there not abundant evidence that now humans are causing changes in centuries that natural processes used do over millions of years?

Well, to answer that question, maybe the wisest course would be to look at an example of climate change in the past that has been widely studied and abundantly documented and see how it compares to what's happening currently. The evidence for this particular climatic event can be found all over the United States and is so obvious that one does not need to be a paleoclimatologist or scientist to recognize the clues and correctly interpret their meaning. One doesn't even have to go out into some remote field area to see the signs. Apartment dwellers and office workers in New York City need only go into Central Park. Armed with no more basic geology than high school earth science, they can see the clear evidence.

In my desk drawer is a piece of wood sealed in a tiny cube of plastic. It looks like ordinary wood, dark brown in color with an appearance of age, but nothing beyond what can be seen in modern forests. It came from a buried spruce tree and was discovered several years ago during foundation excavation for another New York high rise. Such buried trees are not unusual in subsurface excavation beneath the city. Of course, spruce trees don't grow in Central Park today, but everyone knows it's warmer now, so they could have grown there in the past.

True enough, as far as this simple explanation goes, but this little piece of wood has a much more interesting story to tell. It represents a mature forest with giant trees of types that grow today in British Columbia. The old-growth forest had been there for centuries, but during a single instance of time, all the trees were felled somewhere around 13,000 years BP (before present). Beyond telling us something about the climate of the time, these trees present us with a profound mystery of the type that causes people to become scientists. Why was there an old-growth forest where New York City is now located and what happened to it?

The destruction of this forest happened during the current interglacial that began about 18,000 to 19,000 years ago. Typical of an interglacial, the air warmed and ice sheets began a rapid melt back toward the north from the area where New York city is located. South of the glaciers, the tundra followed the ice northward as sea level rose rapidly. Within a few thousand years, the climate in the area had warmed enough to support the spruce forest, but at 12,900 years ago, the deep freeze returned. Known as the Younger Dryas, ice core drilling shows that within only ten years, the climate changed from cool temperate to full glacial conditions. Ten years! The ice advanced south again from the Great Lakes Region and eventually smashed across the spruce forest. The climate warmed again around 11,500 years ago as quickly as it had cooled, returning to the interglacial warming. Once more, the ice resumed its melt back, depositing sediment that buried and preserved the forest.

Inspection of ice core data from Greenland shows that the Younger Dryas was one of the most abrupt changes in climate known.[1] It is detected in several parts of the world, but seems to have been most severe in the northern hemisphere, where the average temperature dropped about 18 degrees F (10 degrees C) in just a few years. The ice core records clearly show that similar abrupt swings in temperature characterize the interglacial from its beginning around 18,000 BP until the warmest part of the interglacial was reached about 7,000 to 8,000 years BP. Climate has continued to fluctuate until the present but the "jiggles" have been much smaller.

The piece of wood and the wide temperature swings of several thousand years ago show how badly climate alarmists exaggerate when they claim we are causing temperature increases in just a few years that used to take millions of years. By global warming alarmists' own admission,[2] warming has amounted to only to 0.75 degrees C (1.35 degrees F) over the past century and a half. Comparing that to the Younger Dryas' 18 degree F swing in ten years answers the question. If they are so poorly informed about something so basic and easily verifiable as the Younger Dryas, how can they claim any credibility at all?

Isn't the ice age a time of truly anomalous climate? As you've said, wide swings of temperature weren't unusual then, so isn't it meaningless to compare what's happening to the climate now with what happened during that period?

The question was about the rates of natural climate change, not about how the Pleistocene ice age relates to other types of climate change. Not even the most fervent global warming alarmist claims humans caused the glaciers to move south, so natural change caused the cold. Under that definition, the cyclical climate swings of the Pleistocene cannot be anomalous if compared to any other type of climate change driven by natural factors. However, the regularity with which the glaciers waxed and waned during that time, while not unique in geologic history, is unusual and deserves more discussion.

People who have heard of the ice age think it ended a long time ago. Many climatologists, however, say that although the earth is in a warm phase, an interglacial, the inevitable swing back toward cold conditions will someday send the ice crashing south again. The reason for such a statement springs directly from a surprising discovery about the Pleistocene from the second half of the twentieth century previously mentioned: the glacial patterns characterizing the ice age were cyclical, with the ice advancing roughly every 100,000 years. This discovery was unexpected and seemed outlandish, but as the evidence accumulated, it had to be accepted. One consequence was the rehabilitation of Milutin Milankovitch, an obscure early twentieth century Serbian meteorologist, but that's getting ahead of the story.

The first clue to the 100,000-cycle of glaciation came from the sea floor and was made possible by a significant advance in paleoclimatology, geochemist Harold Urey's discovery in 1947 that the abundance of oxygen isotopes varies with past climatic conditions.[3]

What are isotopes?

Isotopes are elements that have a different number of neutrons in the nucleus of their atoms forming heavier and lighter forms. Because chemical properties are determined by an element's electron configuration, varying the number of neutrons does not affect them. Among the elements that have isotopes is the oxygen we constantly breathe. Urey discovered that in sea shells, the amount of O^{18} in proportion to O^{16} is higher during colder cycles of climate. In 1955, Urey's graduate student, Cesare Emiliani, applying the technique to tiny shelled marine invertebrates called foraminifera collected from layers of marine sediments, found regular variations in the ratio of O^{18}/O^{16}, which he attributed to changes in ocean temperatures. By 1966, he had worked out a climate curve for the last 425,000 years, which showed the 100,000-year cycle.[4] This went against accepted orthodoxy at the time and became very controversial. Generations of geologists had been trained that there had been only four glacial advances and that there was no particular pattern to them.

His many critics seemed to get the upper hand when they were able to show that O^{18}/O^{16} ratios did not measure water temperature. It wasn't long, however, until other researchers showed that although this ratio didn't measure temperature directly, it was a proxy measure, and that it was a direct measure of the ice volume on the earth, something even more useful for ice age geology.

A direct measure of the amount of ice? How could that be possible?

It's based on the principle that more energy, i.e., higher temperature, is required to evaporate O^{18} because its two extra neutrons makes it heavier than O^{16}; when O^{18}-rich water vapor condenses, it liberates more energy then O^{16}. Because of this, water vapor formed from initial evaporation has more O^{16} compared to O^{18} and the water that's left is the opposite. This reverses when the vapor condenses back into liquid; more O^{18} is taken into the liquid.

Evaporation and condensation are central processes of the hydrologic cycle. Water constantly evaporates from the surface of the oceans and condenses into water vapor in the atmosphere. When precipitation occurs, the process leaves the remaining water vapor with more O^{16}. The end result is that when the climate cools, precipitation that falls is higher in O^{16}. The water vapor that forms during such times will therefore have a lower O^{18}/O^{16} ratio. If the precipitation falls as rain, it will be remixed with surface water. If, however, the water vapor condenses and falls as snow, and is converted into glacier ice, the varying O^{18}/O^{16} ratios are preserved in the layers of ice that form each year until the ice melts.

With the 100,000-year cycle of glaciation well established, work began on trying to figure out what might have caused such a regular and repeated pattern. Within a few years, an obscure theory was pulled out of the dust bin of history, the Milankovitch theory, first proposed in 1920. Milankovitch pointed out in his theory that three recurring variations occur in the earth's orbit around the sun, and that each changes the amount of solar energy reaching the earth's surface.

The first involves changes in the *eccentricity*, the shape of the earth's orbit which varies from nearly circular to more elliptical, with both a 100,000-year and a 400,000-year cycle. When the orbit is more elliptical, the earth is further away the earth is from the sun at it's maximum distance.

The second variation, the angle of tilt of the earth rotational axis--the *obliquity*--causes seasonal change and varies from 21.5 to 24.5 degrees with a 41,000 cycle. The greater the tilt, the greater the difference between summer and winter, with approximately 1% more energy received on the summer hemisphere for each degree of increase.

The third cycle, precession or wobble in the earth's rotation, with a 21,000 to 23,000-year cycle, also affects the seasonal contrast by changing the date of the earth's closest approach to the sun.

Each orbital variation has its own period of several thousand years, and roughly every 100,000 years, the combination of the three results in a deeper minimum of solar energy at the earth's surface, leading to conditions favorable for glaciation.[5] A nice animation showing the three orbital variations is available on YouTube.[6]

This is the best kind of science. A theory that had been considered and abandoned because it didn't explain the facts as they were known in the early twentieth century, was brought back and reconsidered in light of powerful new techniques yielding valuable new data. It is a perfect example of how science is a self-correcting activity and why the scientific method is so powerful.

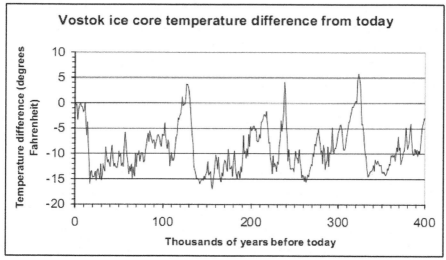

Figure 6. Vostok ice core temperature difference from today

The 100,000-year cycle, initially discovered in sediments on the sea floor, was confirmed by ice core drilling started in the 1970s and 80s. A particularly heroic effort was required to obtain the Vostok ice core from Antarctica in the late 80s. Initially reaching back 160,000 years, it was extended to 400,000 years a decade later, enough to cover the current interglacial as well as the previous three. Figure 6, a graph showing the temperature data from the Vostok core, based on O^{18}/O^{16} ratios in trapped air bubbles, clearly shows the 100,000-year cycle.[7]

Later drilling at other places in Antarctica was able to extend the record back to 800,000 years BP with much greater resolution. Plans are currently underway to try to extend the record back to 1.2 million years. If successful, such a core should be crucial in understanding why a major change in the glacial cycle occurred about 0.9 million years ago, switching from a 41,000-year cycle to the later and better known 100,000 years cycle. This change, still not well understood, was first discovered in deep sea sediment cores extending back 3 million years.[8]

This earlier 41,000-year cycle coincides with the period of one of the three orbital parameters that produces the 100,000-year cycle of glaciation, the earth's axial tilt, which is currently 23.44 degrees, but ranges from 21.5 to 24.5 degrees. It is the axial tilt which causes the

seasons. The smaller the axial tilt is, the less pronounced the seasonal differences. The earth is currently moving in the direction of less tilt, which means sunlight strikes the polar regions less directly, melting less of the winter snowfall, a condition favorable for the growth of glaciers. Up to this point, no explanation for the switch from a 41,000-year cycle to a 100,000-year cycle has been fully convincing.

Does all this have any relevance to what's happening with climate now? Aren't we spewing out so much carbon dioxide that it overwhelms these orbital cycles you talk about, whether 100,000 or 41,000 years?

Drilling into ice caps, whether two-mile thick polar glaciers in

Antarctica and Greenland, or the thinner ice on the tops of high tropical mountains, has great relevance for us. These long straw-shaped samples of ice provide a wealth of information beyond just a year-by-year temperature record. The trapped air bubbles are samples of the atmosphere used to determine the amount of oxygen, carbon dioxide, methane and other gases, and how they varied from year to year. We can tell how much snow fell, the amount of volcanic activity, how dusty the atmosphere was and even learn about the winds. All of this adds tremendously to our knowledge about past climates so we can better understand present changes.

What is most obvious from studying the O^{18}/O^{16} temperature record from the ice cores, such as the Vostok core,[9] (see previous illustration) is just how jagged and irregular the graph is, like a particularly jagged saw blade. There's not a smooth curve to be seen anywhere on the record. Four high, narrow peaks stand out, the warm interglacials. All of them are of much shorter duration than the series of lower jagged peaks that separate them, the cold climates during the glacial advances, each beginning roughly 100,000 years after the previous. The glacial climates were times of much more irregular climate than the interglacials, with times of both global cooling and warming. At least some of these smaller peaks of advancing ice are thought to be related to the periods of individual orbital cycles.

The ice cores show us that over hundreds of thousands of years, the climate of the earth has been variable, but it has been less variable during the warm interglacials. According to the Vostok core, the coldest part of the glacial periods were 15 to 18 degrees C (27-32 degrees F) colder than the interglacials, a huge difference. Compared to such a vast refrigerator, the mild conditions of the current interglacial we now enjoy, which began with rapid warming about 18,000 years BP, is an Eden-like paradise. But even a paradise has its serpents, periods of colder climate, one them being the severe drop 12,900 years ago known as the Younger Dryas.

But you said the Younger Dryas was an exceptional event. Isn't it true that most of the time since the end of the Younger Dryas was characterized by steady climate, without much change, until recently?

Steady, even climate until recently when we began to muck it up, eh? The way to answer that question is to check a few records and see. No matter what time frame is used to look at climate, change is still apparent. For example, if the latest portion of the Vostok ice core record is examined in detail over the last 12,000 years (Figure 7),[10] the now familiar saw-tooth irregular shape is seen, with some prominent peaks and troughs.

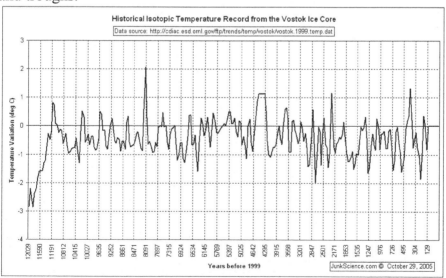

Figure 7. Vostok ice core temperature record, last 12,000 years.

Notice the warmest time in the entire record was about 8,100 years BP, when the temperature was about 2 degrees C (almost 4 degrees F) warmer than today. According to the U.S. Geological Survey, global climate was as much as 4 degrees C (7 degrees F) warmer over the next thousand years, from 8,000 to 7,000 years BP.[11] This period, once known as the Climatic Optimum, before the recent spin that a warming climate is bad news rather than good, and now usually referred to as the Hypisthermal, has possible historic significance. Archaeological evidence suggests that humans first started to give up the hunter-gather lifestyle around this time and settle down in one place to grow crops and domesticate animals as favorable weather spread in the Middle East, where this event seems to have happened first.

Three other peaks in the Vostok record are also warmer than

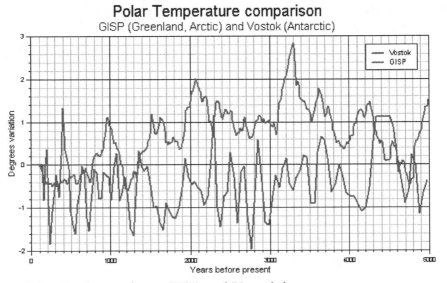

Figure 8. Comparison, GISP and Vostok ice core temperatures.

today, most notably the already mentioned Medieval Warm Period. The coldest time in this record was around 800 BC, followed closely by the Little Ice Age. The Greenland Ice Core record[12] is similar, but differs in details (Figure 8). Looking at this diagram, where the temperature data for the Vostok and Greenland ice cores are plotted on the same graph, makes it clear how much colder Antarctica is (and was).

The warmest part of the Antarctic record peaks later than Greenland, and the Medieval warming is not as pronounced. Both display the jagged up-and-down pattern so characteristic of climate, and both show the recent warming as nothing more than a minor fluctuation.

Many proxy temperature records are available for the most recent few hundred to few thousand years since the Younger Dryas. Tree ring data and various types of sediment, especially fresh water lake and marine, are perhaps the most common tools used to provide estimates of climate over such time scales, but many other proxies have also been tried, some quite novel, such as the layers in stalactites. Results from such studies don't always agree. This is not unexpected considering the variety of techniques used to obtain the data, as well as the fact that certain studies are likely to be of local significance rather than global, such as those involving a single cave or sink hole.

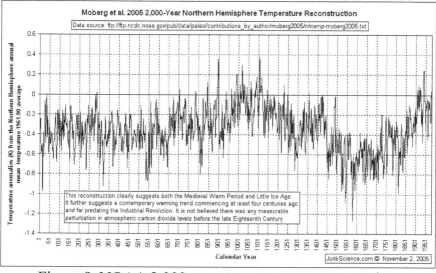

Figure 9. NOAA 2,000-year temperature reconstruction

An excellent example of these shorter term climate estimates, published by NOAA's Paleoclimatology program in 2005, reconstructs temperatures over the last two thousand years (Figure 9).

The graph of temperature,[13] generally in agreement with most studies over a similar time scale, shows the familiar up and down saw toothed pattern of a continually varying global climate. The Medieval Warm Period, the Little Ice Age and the recent warming are the main features. The coldest year on this record is in the late 1500s, with around AD15 being a close second. There is a virtual tie for the warmest year with AD1102 just ahead of AD900.

An especially unusual proxy record of past climates comes from Iceland. Viking ships reached Iceland not long after AD 800, and they soon established colonies. As a sea faring people, they paid close attention to anything that could impede commerce or form a hazard to

57

navigation. Chief among such hazards was floating sea ice that drifted south from the frozen north. The records they kept of the number of weeks out of each year icebergs could be seen floating off the coast of Iceland goes back to the late 800s. A graph of their observations[14] (Figure 10) shows the Medieval Warm Period and the Little Ice Age as clearly as the more recent climate proxies. It is especially

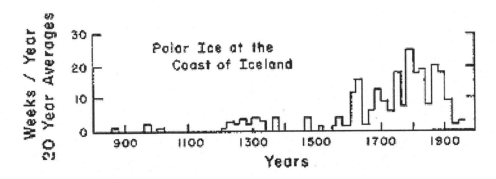

Figure 10. Sea ice floating off the coast of Iceland.

interesting that a historical record matches so closely proxy climate data.

Finally, consider the temperature over the last thirty years,[15] an interval when global warming is supposed to have been rampant. The graph shown as Figure 11 on the may look cluttered at first, but realizing what is being plotted helps to understand it. Temperatures, on the vertical axis, are related to the usual 1951-1980 thirty-year norm, shown as zero on the graph. The squiggly lines plotted in brown (HadCRUT3) and red (GISTEMP) show temperature records from networks of ground-based weather stations, while the one in blue is the satellite record of temperature for the lower troposphere from the University of Alabama-Huntsville (UAH MSU LT). GISTEMP is NASA's record maintained at the Goddard Institute of Space Studies and HadCRUT3 is the record administered by The Met Office in the United Kingdom.

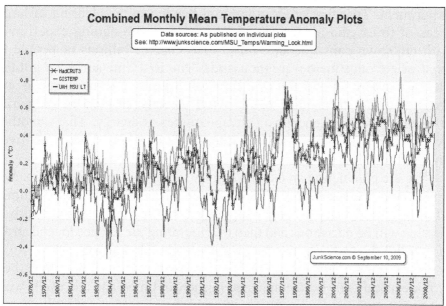

Figure 11. Monthly temperatures over the last thirty years.

The overall pattern looks much the same as other records that show much longer intervals of time. Clearly, climate changes on all time scales, whether long, medium or short. The three plots of temperature agree for the most part, but differ in details. All three show a large spike in temperature for 1998, which, at the time, was widely considered to be proof of runaway global warming, but now known to be related to a major El Niño event at that time. They also agree that since that time, no overall warming has occurred and a small cooling trend has, in fact, developed beginning in the early part of this century.

No one who is worried about climate change has ever denied that climate cycles have operated in the past or that some of the warming in the past 130 years might, in part, be natural. By the same token, climate deniers need to admit that carbon dioxide is a potent greenhouse gas and that artificially dumping billions of tons into the atmosphere every year is a dangerous experiment that has the potential to overwhelm natural cycles.

The only time climate alarmists admit to the possibility that some part of climate change might be natural is when a scientist, who has facts and figures at hand, corners them. As for humans conducting a dangerous experiment by using available and easily obtainable energy resources, by the same logic, it could be argued that everything we've done since first chipping the edge off a piece of flint to make a tool has been an experiment. That first crude stone tool has led to the clearing of large areas of forest and modern agriculture. That's an ongoing experiment with unknown consequences, but it also enables billions of people to exist where once there were thousands. The first time someone got the idea of picking up one end of a limb set afire by a lighting strike and using it to create another fire has led directly to modern industrial civilization, which depends on cheap sources of energy. That's another experiment with unknown consequences, but it led directly to lives immeasurably easier and longer than our distant ancestors.

We are conducting an experiment by using coal and petroleum. We can't know the results of this experiment, because it is still ongoing. Climate alarmists try to scare people by assuming the release of carbon dioxide will be disastrous and then exaggerating the science to lend an air of credibility, shouting that the results are already known

From a geological viewpoint, we can say the experiment with carbon dioxide has already been conducted numerous times, and the results were not disastrous in the past, so why should they be this time? It has been well established that the current amount of carbon dioxide in the atmosphere is much lower than was true for much of the earth's history. As discussed in Chapter Two, life was abundant during those times and there was no runaway greenhouse effect. It is not scientific to assume the results before completing the experiment, particularly when previous runs of the experiment produced different results.

The second part of the comment says that both sides of this debate should agree that adding carbon dioxide into the atmosphere has the power to overwhelm natural cycles. So far, since the nineteenth century, all we've seen is the latest upward tick in temperature after the previous tick in the opposite direction, and even that seems to have reversed in the last few years. Many studies show we haven't even reached the warmth of the Middle Ages yet. Where's the huge leap up in temperature like the one following the Younger Dryas? Show that to people in order to convince them.

If we wait for that, it will be too late to do anything about it. We have to act now, while there's still time. Why isn't the logic of that obvious?

The logic is not obvious, or even sound, because there are too many unknowns. Consider the 70-million year cooling trend beginning in the late Cretaceous Period discussed in Chapter Three. What caused it? We don't know. The leading theory is a change in winds and ocean currents when the plate carrying India began colliding with the Asian plate about

40 million years ago throwing up the Himalayas. The cooling, however, started long before that.

We also don't understand the cyclic nature of the Pleistocene ice advances starting around 2 million years ago, or why the cycle changed 900,000 years ago from 41,000 years to 100,000 years. Beyond that, we don't really understand why the climate got cold enough to send the glaciers south only to eventually warm again. The Milankovitch Theory fits the 100,000 year cyclic pattern fairly well, but seems to fail as an explanation because the orbital variations don't produce sufficient energy decrease from the sun to cause such a severe cooling, only about twice the amount associated with the eleven-year sunspot cycle. Nevertheless, there is a close correspondence between solar energy received and the marine O^{18}/O^{16} ratio raising the possibility that some type of positive feedback mechanism might be involved, something that could amplify a small effect.

This is not new. In fact, Cesare Emiliani suggested such a feedback mechanism back in the 1950s, the Emiliani-Geiss hypothesis.[17] It uses increasing albedo due to increasing ice and snow cover to amplify the small decreases in solar irradiance due to orbital variations. An initial small decrease in solar energy causes the ice and snow cover to expand, increasing albedo and reflecting more energy back into space. This cools the climate even more, causing still more snow and ice, and so on. The effect is reversed when the orbital cycles cause irradiance to increase.

This theory makes sense because we know that changes in albedo have a marked effect on climate. The reader can demonstrate the importance of albedo on surface temperature using a fascinating on-line calculator on the Junk Science web site.[18] It allows changing the value in total solar irradiance, albedo, and greenhouse effect and calculates the resulting average global temperature. According to this calculator, decreasing the irradiance by the maximum due to the orbital cycles, 2.7 w/m^2, lowers global temperature by only 0.25 degrees F; however, increasing albedo from 31 to 32% drops global temperature by almost 2 degrees F.

Indeed, this is the problem for the Emiliani-Geiss hypothesis. Anything that changes albedo has such a powerful effect on the global climate that the small change in solar energy due to orbital cycles would seem to be completely unable to counteract the effect of more snow and ice cover. Some other factor or factors must be operating during the 100,000-year cycle to avoid runaway glaciation. Clouds are a possibility--how many, what type and how often--because they exert such a powerful effect on the earth's albedo as well the absorption of solar irradiance. This relationship will be examined in Chapter Nine.

SUMMARY

Climate alarmists argue that people are now causing changes in the earth's climate in mere centuries that required eons of time in the past. The Younger Dryas, a severe cooling event during the current interglacial that lasted about 1,400 years (from approximately 12,900 to 11,500 years BP) shows how important a little knowledge of earth history can be in the climate controversy. This was a global event because its record is found in both the northern and southern hemisphere. The cold associated with it is clearly seen in the ice core from the Greenland ice cap. Measuring the ratio of O^{18}/O^{16} from trapped air bubbles in the ice, a powerful proxy measure of ancient temperature, shows a sharp drop of about 18 degrees F in only 10 years and an equally speedy warming 1,400 years later. Comparing the 1.3 degrees F of warming during the last 130 years to 18 degrees in 10 years makes one wonder how it is possible for the human-accelerated argument to persist in the media.

Although the study of marine and lake sediments are important in uncovering past climates, our understanding of the last 2 million years of climate history, the time of the Pleistocene Ice Age, was revolutionized by geoscientists Harold Urey and Cesare Emiliani's pioneering work with oxygen isotopes half a century ago. Their discovery that measuring the ratio of oxygen isotopes in marine shells is related to temperature and the waxing and waning of glacial ice showed that there was a very definite cyclical pattern to the Pleistocene: long periods of glaciers advancing far into middle latitudes roughly every 100,000 years separated by shorter intervals, 15,000-30,000 years, of warm climate. Ice core drilling shows that temperatures during past interglacials were warmer than today and that temperatures higher than current also occurred during the present interglacial.

Several examples are discussed illustrating that constant change is central to understanding the earth's climate history. Climates changed in the past, they are changing today and they will change in the future. Climate change has never been smooth, but instead has a pattern resembling fractals. Plots of the temperature display irregular patterns marked by squiggles, saw-toothed patterns of small reversals from a general trend that becomes apparent only in the long term. Both the presence of change and the fractal-like pattern seem to be inherent features of climate change no matter what time scale one chooses to examine.

To illustrate the point, studies of climate change over the past 70 million years, 730,000 years, 12,000 years and 2,000 years are discussed. The warmest climate in the current interglacial was about 7,000 to 8,000 years ago, a climatic optimum that coincides reasonably well with archaeological discoveries showing that human culture began to change at this time from a nomadic hunter-gatherer stage to subsistence farming in permanent settlements. Over the last two thousand years, the warmest time was during the middle ages, as shown by studies in many parts of the world. It was followed beginning in about AD1300 by the cold of the Little Ice Age, ending in the late nineteenth century.

Compared to the past, only minor warming has occurred since then, with the only unusual feature being that many people are now worried they are causing it because burning coal and petroleum emits carbon dioxide. Despite the fact that atmospheric carbon dioxide is low compared to the geologic past and that we have little understanding of the climate system, a crisis has been declared and a call to action sounded.

NOTES AND SOURCES

(1) NOAA's article on the Younger Dryas includes a chart showing that this dramatic cold event is recorded by several different ice cores in in different parts of the world: http://www.ncdc.noaa.gov/paleo/abrupt/data4.html

(2) *Discover*, June, p. 38, 2009, Corey S. Powell (Ed), "The big heat."

(3) *Journal Chemical Society*, Vol. 111, p. 562, 1947, Harold C. Urey, et al, "The thermodynamic properties of isotopic substances."

(4) *Journal of Geology*, Vol. 74, p. 109, 1966, Cesare Emiliani, "Paleotemperature analysis of Caribbean cores."

(5) NASA has nice some nice illustrations of the orbital variations: http://earthobservatory.nasa.gov/Features/Milankovitch/milankovitch_2.php

(6) Animation of the three orbital variations important in the Milankovitch Theory of Glaciation is available on YouTube: http://www.youtube.com/watch?v=wLAYRdSnRSI

(7) A graph of the Vostok ice core temperature data is here: http://aim.hamptonu.edu/outreach/AK-2006/handouts/vostok.pdf

(8) Annual *Reviews of earth and Planetary Sciences*, Vol. 22, p. 353, 1994, M.E. Raymo, "The initiation of northern hemisphere glaciation."

(9) *Ibid*, #7

(10) A nice graph of the last 12,000 years of the Vostok ice core is available on the Junk Science website: http://www.junkscience.com/MSU_Temps/Vostok_short.html

(11) The USGS web site on the glacial history of New York City includes a summary of global climate over the last 20,000 years that includes some information on the Climatic Optimum: http://3dparks.wr.usgs.gov/nyc/morraines/flandrian.htm

(12) Several graphs comparing the Vostok and Greenland ice core data in terms of several parameters is available here: http://mclean.ch/climate/Ice_cores.htm

(13) The abstract and temperature graph of the NOAA study has been published on the web: http://www.junkscience.com/MSU_TemJps/Warming_Proxies.html

(14) A graph of the iceberg data off the coast of Iceland is here: http://www2.sunysuffolk.edu/mandias/lia/determining_climate_record.html

(15) Junkscience.com has a nice graph of temperature over the last several years here: http://www.junkscience.com/MSU_Temps/All_Comp.png

(16) A graph showing the orbital variations, their combined signal and O^{18}/O^{16} ratios is here: http://getenergysmartnow.com/2009/04/23/earth-day-science-climate-and-the-energy-budget/

(17) *Geologische Rundschau*, Vol. 46, p. 576, 1957, C. Emiliani and J. Geiss, "On glaciations and their causes.(18) The JunkScience.com global temperature calculator is here: http://junkscience.com/Greenhouse/earth_temp.html

CHAPTER FIVE—THE GLOBAL WARMING THEORY

Someone said the other day that the global warming theory is controversial because there's not one single piece of evidence to support it. My response was that's wrong because it's a very straight forward theory that has a lot of supporting evidence and is only controversial to right wing Republicans. Who's right?

Ah, the two extreme positions in the climate change debate. While these two views seem to be diametrically opposed, the reality is that they are not. Both statements are, in fact, incorrect, so therefore, neither is right.

How can that be true? Please explain.

Neither of these statements is true because they reflect a lack of understanding as to exactly what the global warming theory is and what it says. There are actually two parts to the global warming theory and only one is controversial.

The first part maintains that starting at some point in the nineteenth century, around 1880 or so, the global temperature began to increase and that during that century and a quarter, it has risen by 1 to 1.5 degrees F. Typically accompanying this declaration is a graph with temperature on the y-axis, the vertical, and time on the horizontal or x-axis, such as the one published annually by NASA's Goddard Institute of Space Studies (GISTEMP),[1] Figure 12, which include a five-year running average to smooth out some of the year-to- year variation.

Other organizations publish similar graphs, which more or less agree with each other (Figure 13).[2]

This part of the global warming theory is uncontroversial, although climatologists do quibble about the reliability of the numbers and whose data set is more reliable. Other than that, climate scientists generally agree that the earth has been warming since the nineteenth century, perhaps beginning in 1850, 1865 or 1880, the most widely cited date because instrument records become more widely available around that time. The widespread controversy comes from the second portion of the global

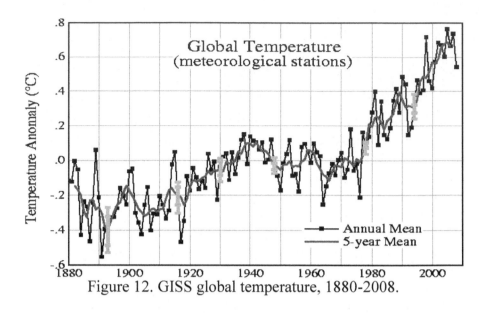

Figure 12. GISS global temperature, 1880-2008.

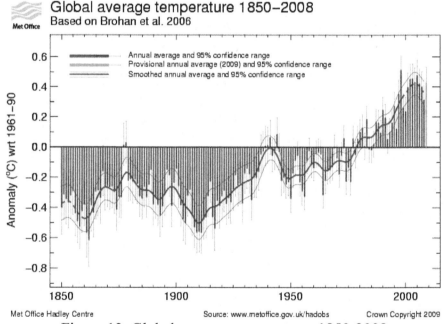

Figure 13. Global average temperature, 1850-2008

warming theory, that people have caused, and are causing, the warming through artificially increasing the greenhouse effect by burning wood, coal and petroleum products. Furthermore, unless action is taken to curb the increase in greenhouse gases, the warming will be a lot worse before this century is over leading to a massive calamity. The basis of this second part of the theory can be traced back to near the time when the warming was just getting started.

As discussed in Chapter One, John Tyndall was the first to show that water vapor and carbon dioxide in the atmosphere slowed the loss of solar heat back into space, as had been first suspected by Joseph Fourier in the 1820s, and compared it to a greenhouse.[3] Swedish chemist Svante August Arrhenius recognized that burning coal and oil released carbon dioxide. In 1896, he calculated that doubling the carbon dioxide in the atmosphere would raise the global temperature by 5 - 6 degrees C (9 - 11 degrees F), a number that agrees well with the published estimates of the UN's Intergovernmental Panel on Climate Change (IPCC) and the forecasts of the most sophisticated computer models. Arrhenius won the Nobel Prize in 1903, not for this work, but for his research on the electrolytic theory of chemical dissociation.[4]

It seems ironic that Arrhenius' simple calculation, which ignored numerous factors now known to be important controls of climate, got nearly the same answer as modern computer projections. It is also perhaps ironic that he is famous today not for his Nobel Prize but for his carbon dioxide work.

One of the few scientists to take Arrhenius' work on carbon dioxide and warming seriously was Englishman G.S. Callendar who, in 1938, argued that carbon dioxide in the atmosphere was already increasing and the global climate was already warming, but the world had more important matters to tend to during that last year of peace. A few other scientists made the same argument during the 1950s, but global cooling was underway then so few paid attention. In 1959, the first observatory to measure and maintain a record of atmospheric carbon dioxide started operating, NOAA's Mauna Loa observatory on top of the highest volcano in the Hawaiian chain. A graph showing the annual trend since then is available on the observatory's web site.[5] During this time, carbon dioxide has increased from about 0.0315 to 0.0388%v (315 to 388 parts per million by volume).

In the 1960s and 1970s, a few scientists continued to sound the alarm over increasing carbon dioxide, but they were mostly ignored because these were years of global cooling. During this time, Emiliani's work on O^{18}/O^{16} and cyclic glaciation due to orbital variations led some paleoclimatologists to predict that the current interglacial was near an end and global cooling would lead to a new ice age. The media, never content to waste a catastrophe, real or potential, began hyping global cooling. Stories and articles with such a theme became common, even in respected magazines such as *National Geographic* and *Popular Science*.

The story *Popular Science* ran in October 1977, "Colder Winters Ahead?" was typical of the time. Accompanying the articles were photographs of glaciers and cars buried in snow drifts.

The climate had already started to turn in the other direction by the time the *Popular Science* story was published, although no one was aware of that yet. Now the time was right because increasing carbon dioxide correlated with a warming climate, and the modern global warming theory came into its own. The visuals were great--time on the horizontal axis, temperature on one vertical axis and carbon dioxide on the other. Just a quick glance made it obvious. We were causing the earth to heat dangerously. Predictions of gloom and doom became common such as Harvard biologist George Wald who is quoted as saying, "Civilization will end within 15 or 30 years unless immediate action is taken against problems facing mankind."[6]

Your treatment of global warming trivializes what is a very important issue. It's a sad fact of modern life that you will be ignored unless the media takes note of you, and to get the media's attention, you have to exaggerate.

On this, we agree. No one feels this more keenly than global warming skeptics. Most of the media routinely ignores everything they have to say. That's bad enough but what is really a sad commentary on the world today is that studies soundly grounded in science that are regularly published in peer-reviewed journals are also ignored in favor of sound bites from such famous and unbiased "scientists" as Al Gore. Nevertheless, shouldn't the exaggerations be exposed and condemned rather than sensationalized and funded?

Consider the GISTEMP record of global temperature since 1880 shown earlier.[7] Notice the last ten years--there's been no increase in global temperature during that time, and since 2001, there's been a cooling trend on the order of perhaps 0.04 degrees F per year. What if I projected this forward 200 years, then called a press conference to announce that the earth will be 8 degrees colder two centuries from now, cold enough to start a new ice age? Wouldn't exaggerating to attract attention justify doing exactly that?

The two are completely different because you don't have any evidence. A few years don't make a trend. All that's happening is a minor fluctuation. Warming will soon be back with a vengeance. On the other hand, there's lots of evidence for global warming, solid, direct evidence.

Well, that's certainly the spin that advocates, the national media and many politicians have fostered. As will be discussed in Chapter Fourteen, they've tried to cut off any critical discussion with their bold assertion that the science is settled and the only thing left to discuss is what we're going to do about it. It might therefore come as a surprise that there are only two lines of direct evidence supporting the second part of the global warming theory, that humans cause the warming. After that, it's all circumstantial.

If that's true, then the two lines of direct evidence must be really solid. What are they?

Actually, according to many scientists, both are shaky.

The first and the most significant of the two is the computer generated predictions, or as climatologists refer to them, General Circulation Models (GCMs). They attempt to calculate how adding carbon dioxide and other greenhouse gases to the atmosphere affects large scale earth systems such as the winds, the oceans, clouds, polar ice caps, albedo, the hydrologic cycle, chemical weathering and many more. In short, GCMs have the ambitious goal of determining the thermal dynamics of the entire atmosphere.

Such a task was unthinkable before the age of supercomputers, but now GCM predictions are treated as if they were real data, not forecasts. It seems that just about everything that's written or said about global warming mentions the forecasts of a much warmer future. They are the foundation that the pyramid of hysteria is built on. Climate alarmists think they are unassailable, a "slam dunk" against which no defense can be mounted. What's important is not what happened in the past, or even present trends (unless it's warming), but predictions for the future. An examination of how GCMs are constructed along with their strengths and weaknesses is saved for Chapter Six.

The other type of direct evidence for the global warming theory is based on laboratory determinations of certain physical parameters, such as the quantity of heat absorbed by a column of air if the proportion of a certain greenhouse gas is increased by a known amount. Such laboratory measured constants are used as the basis for calculating other basic factors essential to constructing the computer models of climate. Foremost among these is radiative "forcing" of temperature from increasing carbon dioxide and other greenhouse gases in the atmosphere. As used by the IPCC, radiative forcing means the average temperature change an externally imposed perturbation, such as increasing carbon dioxide, causes at the earth's surface, as compared to "unperturbed values."[9] While not as well known as the computer projections, radiative forcing is really just as important because the computer models could not be constructed without including a numerical value for this parameter. This physical quantity must be entered into the computer programs if they are to predict how much the earth might warm from increasing carbon dioxide.

It has been a long and controversial struggle, beginning with Arrhenius in 1896, to derive a sound value for what seems like a simple question: if the amount of carbon dioxide in the earth's atmosphere doubles, what effect will this have on the temperature at the earth's surface? The IPCC refers to this basic parameter as "climate sensitivity." If one has a value for radiative forcing, usually given in watts/square meter, climate sensitivity can be calculated.

Certainly, simple radiative forcing can be determined in a laboratory setting where all variables except one, the amount of carbon dioxide or other greenhouse gas, are held constant, but the earth and its many systems are far more complicated than that, so such a number will be of limited value. Even speaking of "the atmosphere" is a gross oversimplification because it is not homogeneous; instead, it consists of various layers (troposphere, stratosphere etc.) with varying amounts of gases in each and a rapidly decreasing atmospheric pressure with altitude.

The real world situation is that if one variable is changed, another one will be affected causing it to change. As the second variable changes, the first one may be affected as well as others. In the parlance of science, this is referred to as "feedback," and it can be positive or negative. Positive feedback means the affected second parameter increases the effect of the first, and negative feedback is the opposite. The latter can be seen to be self-limiting while the former has the potential in theory to behave like a nuclear chain reaction. Because of feedback, one single change in the earth's atmosphere may produce what eventually becomes a complex cascading chain of feedbacks and affected parameters.

The situation grows so complex that obtaining a definitive answer to the question of how much the surface temperature increases from doubling the carbon dioxide in the atmosphere becomes almost impossible to answer definitively. In popular terminology, it's like the "butterfly effect." Will a butterfly flapping its wings in China cause rain next week in New York?

Beginning with Arrhenius, meteorologists and atmospheric physicists have struggled with this question. Results obtained have varied from as little as 0.1 degree[8] to several degrees F. Water vapor and clouds were recognized early as crucial to obtaining a meaningful answer. The calculations gradually became more and more sophisticated as computing power increased the ability to include additional variables and possible feedbacks, but correctly modeling water vapor and clouds remains problematic. A detailed account of the history of these radiative calculations is beyond the scope of the present work, but is available on the web.[10]

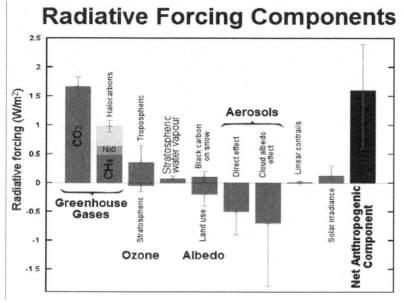

Figure 14. Climate sensitivity, IPCC's Fourth Assessment Report

The most widely cited value for climate sensitivity comes from the IPCC's Fourth Assessment Report (Figure 14)[11] which estimated 3 degrees C (5.4 degrees F). Widely different estimates of climate sensitivity, however, have been obtained, ranging from less than one Figure 14. Climate sensitivity, IPCC's Fourth Assessment Report to greater than ten degrees.[12] The IPCC report also included a widely reproduced graph,[13] summing the positive and negative forcing agents, i.e., those that cause warming such as greenhouse gases and those that result in cooling, mainly sulfate aerosols. The IPCC's estimate, using -0.6 W/m² for aerosols, of a total anthropogenic positive forcing of 1.6 W/m², is the basis for their prediction of warming climate in the future.

An obvious problem with this estimate is that the amount of radiative forcing from anthropogenic aerosols is not well known. Estimates have ranged over a factor of ten, from -0.3 to -3.0 W/m². Combined with a similar range of uncertainty for climate sensitivity, it can readily be understood why estimates of the total global mean present-day anthropogenic forcing range from +3 W/m² to -1 W/m².[14]

That's a huge range, all the way from warming to cooling. With such disagreement about the most basic parameters used in the GCM forecasts, it's no wonder global warming skeptics argue that GCM forecasts pointing to a certain number of degrees warmer climate by 2100 rest on a foundation of Jell-O.

While it's true the uncertainty in determining climate sensitivity is greater than we'd like, it's not as great as you make it sound, and the models are improving all the time. Even without the general circulation models, there's a lot more evidence. Some are things we can already see happening, early signs of what is to come. As a scientist, don't you agree that the strength of a theory is measured by the success of its predictions? And if we can see predictions of the global warming theory starting to happen, there is good reason to think it is correct, isn't there?

This view of scientific theories is right, but incomplete. In order for a scientific theory to become widely accepted, it must have two properties. First, as stated, it must make predictions. But not just any predictions. To be a real scientific theory, there must be some way to test the predictions, and the tests cannot be rigged, they must be such that the test can be failed. In other words, the predictions a theory makes can turn out to be wrong.

The second property necessary for a robust scientific theory is that all predictions the theory makes must be successful. Just one failed prediction, and the theory has been shown to be wrong. This is the falsification model of science that is discussed more fully in Chapter Fourteen. A falsified theory is one that fails at least one prediction. In an ideal world, it should be abandoned after it has been falsified, i.e., proven wrong in at least one prediction.

The global warming theory does make predictions. Many of them are conceivably testable. That's not to say all the tests are feasible, but that's not a requirement. In principle, human-caused global warming is a testable theory capable of being falsified, a genuine scientific theory.

Some scientists argue that the predictions of the theory, or at least one prediction, has already failed. Others say more time is needed before we can be sure about this.

The time objection is well known. Keep putting it off until we're sure. It's just a lame excuse for doing nothing. I'm more interested in what you call failed predictions. Can you give an example?

A number of scary books about global warming started being published twenty or twenty-five years ago. This was a time when the liberal media had not yet enlisted in the climate change legions, and these books did a lot to convince them. Their authors knew they were exaggerating, but they did it to draw attention to what they fervently believed was a worthy cause. Typical of these early global warming books, *Dead Heat: the Race Against Global Warming* by Michael Oppenheimer and Robert Boyle was published in 1990 warning of coming disasters if we didn't act quickly. Of course, global warming alarmists still sing the same verse. The book is full of gloom and doom scenarios awaiting us a century hence, of the same kind found in Al Gore's book and movie *An inconvenient Truth.* These are in principle testable predictions, if we wait a few generations.

One thing about *Dead Heat* is really interesting though. In addition to the long term forecasts, the authors also make predictions for ten, twenty, thirty years. That's a bold thing to do, a dangerous thing to do, because people live long enough to check on them. It's these short term predictions that are so interesting to read, not just because each of them is so far off the mark, but because of the general tone. It's hard to imagine that someone writing in the late 1980s would actually think conditions could get so bad so quickly. Already warming is supposed to be so great that crops failures are widespread resulting in large scale food riots and the overthrow of governments.

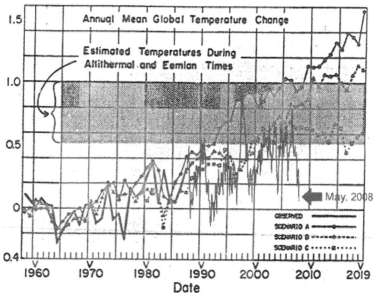

Figure 15. Hansen's predicted temperatures from 1988

At the time *Dead Heat* was published, general circulation models were already being highly touted. James Hansen of NASA's Goddard Institute of Space Studies used a chart of a GCM to illustrate his speech to Congress in 1988 that was widely publicized at the time (Figure 15).

It predicted global temperature out to 2019 according to three scenarios. Scenario A was business as usual, doing nothing to curb greenhouse gas emissions and producing the greatest increase in global temperature. Scenario B was doing a little and C involved taking immediate and stringent measures to reduce carbon dioxide emissions, thereby producing the least temperature increase. His Scenario A is the one the world has actually followed. Comparing the predictions to actual temperature data since then shows that all three scenarios greatly over estimated global temperature, with Scenario A showing that already it should be hotter than peaks reached in previous interglacial periods (shown in the graph as Altithermal and Eemian Times). Comparing the predictions from his chart to present temperatures, plotted on the graph in red,[15] clearly illustrates how wrong these forecasts actually are.

If this is not falsification of the global warming theory, it's getting uncomfortably close to it. Certainly, it shows the GCM forecasts from the late 1980s were incorrect, and since they were considered the strongest bit of direct evidence for global warming at that time, just as they are today, the theory itself becomes questionable.

Maybe so, for GCMs as they then existed, a generation ago, but computing power has increased greatly since then and general circulation models have gotten a whole lot more sophisticated. Isn't that correct?

There's no doubt of that, but does that mean the GCM forecasts have improved? Check the chapter in *Dead Heat* on general circulation models and compare the forecast to more recent examples. A good example is the mean of ten recent GCM forecasts published in a peer-reviewed journal[16], or the IPCC's projection of 21st century global temperatures shown as Figure 16.[17] Differences between the predictions of older and newer GCMs are insignificant.

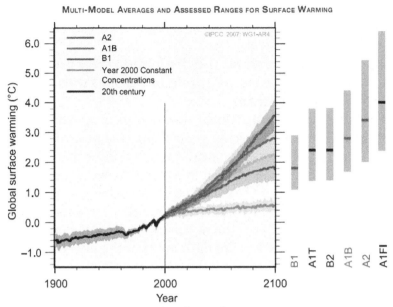

Figure 16, IPCC's projection of 21st century temperature

Certainly general circulation modes are able to consider many more factors related to climate than in the past, but they are still crude compared to the many systems that potentially affect the earth's climate system. Scientists who construct climate models readily admit the deficiencies in peer reviewed journal articles, but these caveats somehow are forgotten in magazine articles and television programs. This might be due to the media's attempt to sensationalize coverage. It could also be due to a scientist slipping from being an objective observer to an advocate for a cause. Consider that Stephen Schneider of Stanford University has been in the climate modeling business about as long as anyone. He was famously quoted in an October 1989 interview in *Discover* as saying that each scientist, "... has to decide what the right balance is between being effective and being honest.[18] Dr. Schneider maintains it is misleading to take only a portion of his remarks on this subject out of context, so a link to the complete quotation is provided.[19]

There's a lot more evidence for global warming then physical constants measured in a lab and computer climate models. How about melting glaciers, rising sea levels, more hurricanes and hotter summers? We see these things happening, just as the global warming theory says they will. What about all this other evidence?

Of course, a lot more evidence is offered as proof of global warming than physical quantities and GCMs. As mentioned earlier, all of these predicted effects are indirect.

I'm not sure I understand the distinction between direct and indirect effects or evidence, as applied to global warming. Could you clarify what you mean?

Recall the two parts to the global warming theory: that the earth has warmed since the nineteenth century and that humans are causing it. Only the second part of the theory is controversial. Evidence that relates to that portion is direct evidence. The other things mentioned, plus a lot more, if they were happening, might be evidence for warming, but not that people are causing the warming. The only real evidence offered that humans are to blame is the simulations of climate run on supercomputers. Because of their special importance, they are the subject of the next chapter.

Subsequent chapters discuss all the more important pieces of indirect evidence. As a preview, consider an interesting web site that relates to one of the things mentioned a moment ago, hotter summers.

If summers are really getting hotter as we hear all the time, we would expect to see more record high temperatures being set. This is the reasoning behind the animated map of the United States (lower 48 states) showing state-by-state record high temperatures for each decade beginning in the 1880s. Click on the map[20] and let the animation run a few times. See if there is a pattern. It's clear that the 1930s was the hottest decade in this time period in terms of the number of record high temperatures.

SUMMARY

Critics argue that the global warming theory has no supporting evidence, but that is incorrect and reflective of a misunderstanding of the theory. In reality, the theory of human-caused global warming consists of two parts, with only one of them being controversial. The first portion of the theory, the uncontroversial part, is that the earth began to warm at some point in the nineteenth century, perhaps 1850, 1865 or 1880. Although there is disagreement over how much warming has occurred, most scientists accept this as fact.

The controversy arises from the second part of the theory which states that human activities, particularly burning wood, coal and petroleum products to produce energy and releasing carbon dioxide, has caused and continues to cause the warming. This idea can be traced back to Swedish physicist Svante Arrhenius in the nineteenth century. Based on laboratory-obtained measurements, his calculations showed that doubling the amount of carbon dioxide in the atmosphere would produce 4-5 degrees C (7-9 degrees F) of global warming.

Both direct and indirect evidence is argued in support of the much debated second portion of the global warming theory. Direct evidence is concerned with modern civilization and industrial activities being the immediate cause of a warmer climate. In contrast, indirect evidence deals with the predicted, supposed or imagined consequences of a warming climate, rather than the cause or causes.

The most widely cited direct evidence that humans cause global warming is what climatologists refer to as a general circulation model (GCM). These are the computer-generated predictions of many degrees of warming that is supposed to occur by the end of this century. Although the models have grown far more sophisticated with increasing computer power, the amount of warming they predict has not changed that much from the 4-5 degrees that Arrhenius calculated.

Computer models of the earth's climatic systems depend on a foundation of laboratory-determined values for how various greenhouse gases affect surface temperatures. Experiments in which such physical radiative values are measured might also be thought of as direct evidence. Two quantities, radiative forcing and climate sensitivity, are widely referenced in the global warming debate, so it is important to understand them.

Radiative forcing, as the IPCC defines it, refers to the amount of mean temperature change increasing carbon dioxide and other greenhouse gases cause at the earth's surface. Forcing is usually measured in watts per meter squared (W/m^2). Climate sensitivity, calculated from radiative forcing, refers to how much global temperature might increase from a doubling of carbon dioxide. While measuring radiative forcing in a lab by adding carbon dioxide to a column of air is a straight-forward procedure, the actual climate system is far more complicated because of the many possible feedbacks. Positive feedback increases the effect of some factor, producing more of it, but negative feedback can be self-limiting because the initial factor is decreased. Because estimates of radiative forcing have varied over a factor of ten, GCM forecasts are a lot less certain than usually presented. In cases where their accuracy can be checked, such as predictions made in the 1980s for the present time, they are proven to be wildly off the mark.

NOTES AND SOURCES

(1) For a nice graph of the GISTEMP record, see: http://data.giss.nasa.gov/gistemp/graphs/Fig.A.lrg.gif

(2) Compare GISTEMP to the one published by the United Kingdom's Met Office: http://hadobs.metoffice.com/hadcrut3/diagnostics/global/nh+sh/

(3)Wikipedia's biography of Tyndall is at this link: http://en.wikipedia.org/wiki/John_Tyndall#Main_scientific_work

(4) A short biography of Arrhenius is at: http://chem.ch.huji.ac.il/history/arrhenius.htm

(5) For a graph of the Mauna Loa observatory data, see: http://www.esrl.noaa.gov/gmd/ccgg/trends/co2_data_mlo.html

(6) This and many other wrong earth Day 1970 predictions are available at the following web site: http://www.ihatethemedia.com/earth-day-predictions-of-1970-the-reason-you-should-not-believe-earth-day-predictions-of-2009

(7) *Ibid*, #1.

(8) *Science*, Vol. 207, p. 1462, 1980, Sherwood B. Idso, "The climatological significance of a doubling of earth's atmospheric carbon dioxide concentration."

(9) The IPCC's Third Assessment Report is available on the web: http://www.grida.no/publications/other/ipcc_tar/?src=/climate/ipcc_tar/wg1/212.htm

(10) For a really well done history basic radiation calculations, see: http://www.aip.org/history/climate/Radmath.htm

(11) The Fourth IPCC Assessment Report is available here: http://www.ipcc.ch/ipccreports/ar4-wg1.htm

(12) *Journal of Geophysical Research*, Vol. 106, p. 5279, 2001, S.J. Ghan et al, "A physically based estimate of radiative forcing by anthropogenic sulfate aerosol."

(13) The IPCC graph showing radiative forcing components can be viewed here: http://en.wikipedia.org/wiki/File:Radiative-forcings.svg

(14) *Ibid*, #12

(15) The chart Dr. Hansen used in his speech to Congress in 1988 is reproduced at this site: http://www.climate-skeptic.com/2008/06/great-moments-in.html. A link at this site is provided to Hansen's text and chart.

(16)*Climate Dynamics*, Vol. 17, p. 83, 2001, S.J. Lambert & G.L. Boer, "DMIP1 evaluation & intercomparison of coupled climate models."

(17) *IPCC 2007 Summary for Policymakers*, Figure SPM-5 of the Working Group I "Summary for Policy Makers," available on the web at: http://ipcc-wg1.ucar.edu/wg1/wg1-report.html

(18) *Discover*, p. 45, Interview of Stephen Schneider, October 1989, J. Schell.

(19) Stephen Schneider's full remarks on making scary statements for the media are available at this web site: http://theresilientearth.com/?q=content/airborne-bacteria-discredit-climate-modeling-dogma

(20) The animated map showing high temperature records for each state in the "lower 48" is available at this web site: http://hallofrecord.blogspot.com/2009/01/decadal-occurrences-of-statewide.html

CHAPTER SIX—COMPUTER PROJECTIONS

What is the difference between a computer projection and a prediction?

The terminology used in association with the computer programs known as general circulation models or GCMs (introduced in Chapter Five) can be confusing. They are also sometimes referred to as "climate models," "computer models," or "general climate models." All refer to the same thing: incredibly complex computer programs that attempt to simulate the climate of the earth at various points in the future. The programs are so long and involved that they could take a lifetime to produce a prediction on a desktop computer, which is why they are run on the world's fastest supercomputers.

Two words should always be kept in mind when considering GCMs, simulations and models. Both refer to something that is like or resembles something else, but it is not the real thing.

Many dictionaries list "making a prediction" as one of the meanings for the word "projection," so in the dictionary sense, the two words are synonymous. From a practical standpoint, however, projection has taken on a higher degree of certainty than prediction, even among climate scientists. A recent survey of climate scientists found that 20% associate *possible* with prediction, but 29% associate *probable* with projection.[1]

This is in no small measure due to way the IPCC and the media portray the forecasts of warmer climate the various GCMs continue to churn out. As one of the two types of direct evidence supporting the global warming theory, since 1995 the IPCC has treated them as increasingly certain, gradually elevating them to their current superstar status. Because they are products of computers, they are treated as if they were unbiased. People make mistakes, but not computers, so they must be accurate and real. People forget that they involve things that *might* happen at some point in the future, and that they are based on scientists' understanding of all the underlying physical, chemical and biologic processes that control the earth's climate, as well as all the various feedback processes that play such crucial roles. If we had a full understanding of all these controls and were able to portray them accurately and precisely in a computer program, a reliable prediction might be possible. Of course, that's far from the case, even for such a basic physical parameter as what value to use for radiative forcing, as Chapter Five explained.

Beyond just the physical parameters, don't the computer climate models also incorporate data from past climate change in their forecasts?

Referring to the outcome of the computer programs as projections, one might think so since at least one thesaurus mentions "extrapolating from past observations" in reference to the word "projection."[2] Climate modelers can't do this, however, because they don't understand the causes of past climate change. They don't even use the most recent records of global temperature over the past several years. Tapio Schneider, a well-known environmental writer admits this: "However, climate models are not empirical, based on correlations in such records, but incorporate our best understanding of the physical, chemical, and biological processes being modeled."[3]

To a geologist, it seems highly presumptuous to think one can make a reliable prediction concerning future climate without first having a well-grounded understanding of the causes of past climate changes. This is especially true now because the past 2 - 2.5 million years, the Pleistocene Ice Age, was not a time of normal climate. From a geological perspective, we are living in what is only the latest interglacial in a long series that came every 100,000 years, and earlier, every 41,000 years. Comparing the current interglacial that started about 18,000 years ago to how long previous interglacials lasted suggests that its end may not lie too far in the future. It should seem obvious that attempts to model climate and predict future conditions cannot succeed without a firm and full understanding of the natural trends and causes of past climate change. Whatever changes humans might cause will be superimposed on natural controls and the interactions and feedbacks between them are likely to be complex. To focus on the human component without being able to explain the glacial cycles and other past changes in climate is not a procedure likely to produce reliable climate forecasts.

For climate modelers to present computer predictions to government leaders as reliable forecasts of the world's climate in coming decades requires abandonment of science in favor of advocacy. Global warming skeptics are near unanimous in charging the IPCC with exactly that. Of course, the IPCC insulates itself from such charges with statements such as the following from a 2001 report: "In climate research and modeling, we should recognize that we are dealing with a coupled non-linear chaotic system, and therefore that the long-term prediction of future climate states is not possible."[4] Of course, this doesn't stop the IPCC from highlighting graphs in their reports, and especially in the summaries they prepare for policy makers, that do exactly that.

When was such a qualifying statement ever highlighted in an article or news report of the latest IPCC projection of a much warmer earth? To my knowledge, never. Instead, these computer predictions are treated as they were scientific fact. Typically, the most severe of the warming scenarios are singled out and discussed. News organizations must be completely unaware of the following admission from one the author's of the IPCC Third Assessment Report: "In fact there are no predictions by IPCC at all. And there never have been. The IPCC instead proffers "what if" projections of future climate that correspond to certain emissions scenarios. There are a number of assumptions that go into these emissions scenarios. They are intended to cover a range of possible self-consistent 'story lines' that then provide decision makers with information about which paths might be more desirable. But they do not consider many things such as the recovery of the ozone layer, for instance, or observed trends in forcing agents."[5]

I'm more interested in the graphs showing future climate that the IPCC publishes. Don't they show the amount of error that goes into constructing them? And isn't that so small that it doesn't make much difference? That seems scientific to me, so what is there to complain about?

That's right, they do include the error bars so characteristic of scientific reports. For instance, the most recent of such graphs, one that the left-leaning media has widely reproduced, is Figure SPM 5 on page 14 of the IPCC's 2007 Report for Policymakers,[6] showing surface warming to 2100 relative to 1980-1999, shown in the diagram as zero. This IPCC graph is reproduced as Figure 17. Four scenarios of warming are depicted in various colors with B1 being the least severe (carbon dioxide reaching 600 ppmv) to A2, the most severe (carbon dioxide reaching 1250 ppmv.) For contrast, an orange line is also shown where carbon dioxide is held at year 2000 levels, but it is not included as one of their scenarios. These shaded regions show the amount of error, and they make little difference. According to these diagrams, there's a lot of warming

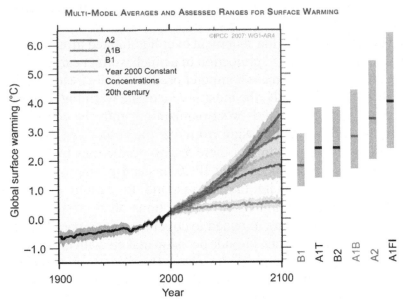

Figure 17. IPCC's projections for temperature. 2001-2100

in the future. Anyone thumbing through this report and looking at this graph can't help reaching that conclusion. Of course, that's the IPCC's intention, for this diagram as well as the entire report.

But what type of error does the shaded region show? The caption for Figure SPM.5 explains: "Shading denotes the +/- standard deviation range of individual model annual averages." What does that mean? Standard deviation is a statistical measure of variability.[7] A low value means that the data points are close to each other, while a high value refers to more scatter. The term is also used to indicate confidence in statistical conclusions, or amount of error, with one standard deviation equaling 68% confidence and two standard deviations equaling 95% confidence, the margin usually reported in scientific studies.

The shaded region on the IPCC graph therefore is for a 68% confidence level or margin of error, but what type of error does this refer to? One would naturally think it means the warming shown out to the year 2100 has more than a 2/3 chance of actually happening. Of course, that's the intended conclusion.

You're saying that's not what it shows? What else could it mean?

The caption for the graph explains this in specific language. "Shading denotes the +/- standard deviation range of individual model annual averages." As pointed out by Ph.D. chemist Patrick Frank,[8] this means

82

that the result obtained by the GCMs falls in the shaded region 68% of the time. It is stating, in effect, what the precision of the model is, but it is silent on the accuracy. In other words, it says nothing at all about the many uncertainties of poorly known physical quantities that go into constructing these general circulation models.

I'm not sure I get the point. One standard deviation, 68%, doesn't seem bad for nearly a hundred years in the future.

Certainly not, but that's not what the shaded areas of the graph show. An example, modified from Frank,[9] makes this clear. What if a research group wrote a computer program that said the average person's height is ten feet +/- the standard deviation range of two inches? This means that every time they ran the program, the result would be within that two-inch range. A graph showing the result would have a shaded region two inches wide, quite a precise result. If two dozen other research groups had computer programs that gave similar results, a press conference might be called to announce that scientists have reached a consensus on the question of human height.

A few years later, after a lot of government grant money and additional work, the same groups announce that the computer programs are now even more sophisticated so the results are +/- 0.2 inch. Widespread media coverage highlights that now the computer programs are ten times better.

Of course, the models are just as wrong as before, or inaccurate, but they are more precise, meaning the wrong numbers they calculate are closer to each other. This is the critical difference between accuracy and precision, something the IPCC reports never mentions.

That's a ridiculous argument fallacy, a reduction to absurdity. You don't actually claim the IPCC graph is as inaccurate as your example, do you?

Scientists' uncertainty of the physical controls of climate that go into GCMs is far greater than people realize. For instance, the peer-reviewed journal *Climate Research* published an article in 2001 showing just how large the uncertainties of the physics really is.[10] One will look in vain for any such admission in the IPCC's 2007 Summary for Policymakers or in the Technical Summary of the latest Assessment Report, the Fourth, released in 2007. One must turn to the supplemental materials accompanying Chapter Seven, "Climate Models" for graphs displaying such data. There graphs show individual GCM errors for shortwave radiation reflected to space ($\sim$25 W/m^2), outgoing longwave radiation ($\sim$20 W/m^2) and surface heat flux into the oceans ($\sim$10 W/m^2),[11] One wonders why the graphs showing such huge errors were not placed in the shorter summaries where they were much more likely to be seen by government officials. Could it be that the IPCC did not want policymakers to know that the scenarios predicting future warming are nothing but stories?

Maybe they didn't think they mattered enough to policymakers to waste their time on them. The important thing is the graph projecting warming, isn't that so? That makes the danger clear.

Yes, the graph certainly does that, but only if one does not understand the importance of the errors. Consider just one of the uncertainties, the 25 W/m^2 in shortwave radiation reflected into space. As mentioned in Chapter Five, 1.6 W/m^2 is the total estimated positive radiative forcing for all anthropogenic greenhouse gas emissions.[12] Now consider that the uncertainty from just one physical parameter is more than 15 times that. There are similar uncertainties for outgoing longwave radiation and surface heat flux. How then, can someone claim to predict something, such as warming, if the error involved in the procedure is more than 15 times greater than the effect one is trying to predict? Another imponderable is why the IPCC puts a graph in their only short, easy-to-read and hard hitting publication, the Report for Policymakers, that contains a graph labeled with one standard deviation error, when the uncertainties in GCMs are far greater than the effects they're claiming to predict.

The media treat the projected warming as reliable and so do most people in government, yet you claim they're just the opposite. You can understand my confusion. Is there any way to actually test these models?

Well, of course one way is wait around and see how well the IPCC prediction did compared to actual temperatures at the time, but activists are clamoring for action now, so that's not very satisfying. Besides, a generation or two from now, people will be concerned with the IPCC's current predictions, not one made fifty years ago. However, climate alarmists have been making predictions of future calamity from a warming earth long enough to see how well they've done. It probably won't be a surprise that they have fared poorly. As mentioned in Chapter Five, James Hansen of NASA's Goddard Institute of Space Studies

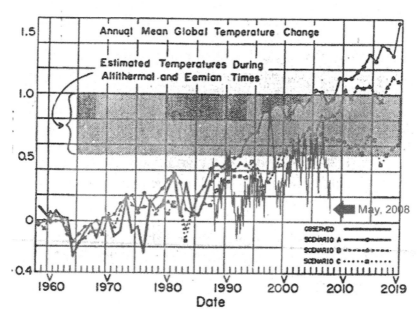

Figure 18. Hansen's predicted temperatures from 1988

used a graph in his 1988 testimony to Congress very much like the one in the IPCC 2007 Report for Policy Makers. He showed three scenarios of warming, with Scenario A, doing nothing, the scenario nations of the earth have actually followed because atmospheric carbon dioxide has continued to increase every year since then. Perhaps the earth didn't get the message because, according to Hansen's chart,[13] global temperature should already be about 2 degrees C (3.5 degrees F) warmer than it is today (Figure 18). Even his Scenario C, taking stringent action to curb carbon dioxide emissions, overestimates current global temperature by about one degree F.

The 1990 best seller *Dead Heat*, in some ways, an earlier version of Al Gore's book and movie, overestimates current global average temperature just as badly as Dr. Hansen.[14] "By then (early in the twenty-first century), the world will be about a degree warmer than it is now." One degree is more than ten times the actual change since the book was published. In referring to climate model predictions, the authors say that by 2020 to 2030, "the mean temperature of the planet can be expected to reach six or seven degrees above the current level."

Another approach to gauging the validity of GCM forecasts is to see how they simulate reality in specific elements of climate, rather than something as broad as average global temperature. One key component is their ability to reproduce the actual vertical structure of the troposphere, the lower layer of the atmosphere where all the earth's weather originates. A 2004 study in the peer-reviewed journal *Climate Research*[15] investigated this by comparing temperature trends in the troposphere from 1979 - 2000 with a series of climate simulations. Being able to reproduce this is especially important because a fundamental feature of general circulation models is that they show more warming in the troposphere than at the surface. The results (Figure 19) showed none of the models studied were able to reproduce the actual tropospheric temperature trends. The authors noted that failure of the GCMs in this respect suggested, "...caution in applying simulation results to future climate-change assessment..."

PROJECTIONS OF SURFACE TEMPERATURES

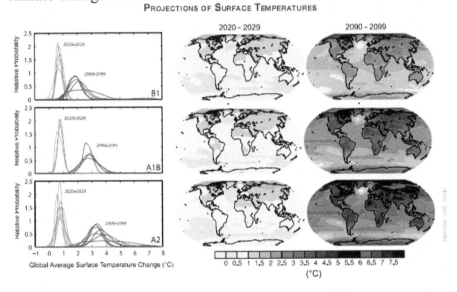

Figure 19. Projected troposphere temperature (right) compared to actual Troposphere temperature (left)

Another feature common to all general circulation models is that they portray greenhouse forcing as producing varying amounts of warming over the earth's surface, rather than uniform warming. Polar regions are supposed to amplify the warming. In other words, the Arctic and Antarctic are supposed to warm the most, equatorial areas the least, with mid-latitude areas being in between. Figure SPM.6 in the IPCC's latest Summary for Policymakers shows this clearly,[16] and it has been a consistent feature of GCM forecasts for many years. Therefore, a particularly significant test of whether GCMs are likely to be reliable would be to compare temperature records from some area in the high arctic with the predictions. It is unfortunate that few long-range arctic instrument records exist, because such high latitudes are supposed to be where the fingerprints of global warming first appear.

In this regard, a peer-reviewed study published in *Geophysical Research Letters* in 2002 is especially valuable.[17] The study was made possible by a newly released set of thermometer temperature records, sea-ice extent and ice thickness from coastal weather stations in eastern Siberia just south of the Arctic Circle extending back 125 years. The authors found that the temperature trends did, "not support amplified warming in polar regions predicted by GCMs." It's a significant study because observation data is not in agreement with predicted warming.

Although the instrument record covers only 1986-2000, a study published in the peer-reviewed journal *Nature*[18] found that the McMurdo Dry valleys of mainland Antarctica cooled over that fourteen year period at the dramatic rate of almost 1.3 degrees F per decade. The authors note the rapid cooling trend they found agrees with a 35-year cooling trend found in other areas of mainland Antarctica.

Cooling? I thought Antarctica was warming, that the ice was melting.

One would certainly think so from the never-ending onslaught of media hype and sensationalism, but the reality is different. Only the Antarctic Peninsular is warming, just a small part of the continent. Chapter Ten will examine the Antarctic situation in more detail, especially what's happening with the ice. Incidentally, the McMurdo Dry valley study found that the ice on the always ice-covered lakes thickened almost five feet during the fourteen year period of their study.

A 2004 peer-reviewed study in the Journal of Geophysical Research[19] focused on how well various GCMs were able to simulate present-day climate in Europe as well as during the middle part of the current interglacial and during the latest glacial maximum. Taken as a whole, the GCM faired poorly with a lot of "inter-model variability." Perhaps the most interesting result of the study is that for the last glacial maximum, the scientists found that when they substituted in the models newly available data on the oceans, the temperature performance of the GCMs deteriorated.

A final real world test of general circulation models involves their ability to predict how doubling carbon dioxide in the atmosphere might effect the amplitude and frequency of El Niño-Southern Oscillations (ENSO). A 2006 peer-reviewed study used the same 15 GCMs cited in the IPCC Fourth Assessment Report.[19] The study found, "Under carbon dioxide doubling, 8 of the 15 models exhibit ENSO amplitude changes that significantly exceed centennial time scale variability within the respective control runs. However, in five of these models the amplitude decreases whereas in three it increases: hence there is no consensus as to the sign of change." In other words, seven of the GCMs predicted no significant change in ENSO, five forecast weaker El Niños and three, stronger El Niños.

What does that tell us about El Niños in a carbon dioxide enhanced world? Nothing, but it does speak powerfully about the limited forecasting ability of GCMs.

Maybe the models aren't perfect, but they're the best tool we have, and they're getting better all the time, more reliable and better able to simulate the real world climate system. Isn't this correct?

Certainly that's what the modelers say and it's their justification for more and more grant money to buy more gigazillion-dollar supercomputers. But whether bigger is actually better is open to question. In a 2008 article,[20] chemist Patrick Frank devised an arithmetical model for predicting greenhouse warming so simple the calculations can easily be done using a hand calculator:

Global Warming = 0.36x (33 degrees C)x[(Total Forcing)/(Base Forcing)]

All the terms Frank used in the equation were taken from peer-reviewed scientific studies and references to each of them are listed in his article. The "0.36" refers to the proportion of greenhouse warming attributed to carbon dioxide plus a supposed water vapor feedback, which will be discussed in Chapter Nine. The second term, "33 degrees C," is derived by comparing the amount of energy the earth emits into space with the mean surface temperature in 1900. Frank calculated the numerical values used in the final two terms, "Total Forcing" and "Base Forcing," using IPCC-approved methodology.[21]

To test the physical accuracy of his simple passive model, Frank compared it to the temperature predictions of the ten advanced GCMs the Lawrence Livermore National Laboratory studied in their "Coupled Model Intercomparison."[22] His Figure 2a, shown as Figure 20, plots the predictions of the ten GCMs out to 80 years along with a line showing the average prediction of the ten climate models and another line showing the result of his simple mathematical model.[23]

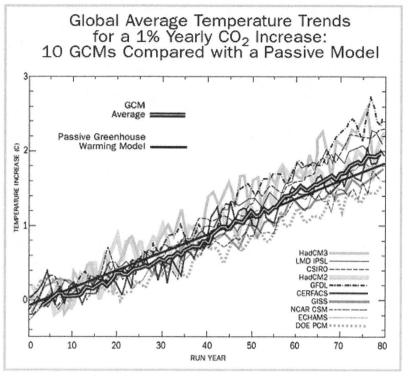

Figure 20, 10 GCMs compare to a simple passive model

Climate modelers generally consider the average of several GCM predictions to be more physically accurate than the output of single models.[24] An inspection of Frank's Figure 2a[25] clearly shows that by this criterion, his own simple model is more accurate than any of the ten tested climate models because its output is closer to the average than any of the tested GCMs. For all their multimillion dollar cost, the complex and supposedly sophisticated general circulation models do nothing more than simply respond in a linear manner, "to greenhouse gas forcing."[26]

Scientists who actually work with general circulation models state that a good test of their accuracy is to see if they can reproduce past climates. One of them recently said, "When we entered into the computer all the various things that forced the climate to change, we were able to faithfully reproduce the temperature record of the past 100 years globally. When you take out the component of human-generated carbon dioxide, the models don't work at all."[27]

This is one of the favorite arguments of climate modelers because it is so convincing to the general public. If a model can "postdict" the temperature record for the past 100 years by including our carbon dioxide addition, but not if that's left out, most people see that as really strong evidence that the prediction for the next 100 years must be accurate. Modelers have been using this argument for a long time, but they keep using it because it sounds so good.

For instance, James Hansen of the Goddard Institute of Space Studies published an article in a 1981 issue of the journal *Science* that made this claim,[28] and nearly twenty years later, P.A. Stott published another report of research in the same journal that used the same argument.[29] The argument sounds powerful until one understands the huge amount of uncertainly that goes into the variables that are used in programming climate models. All the modeler has to do is to choose from a range of values for carbon dioxide forcing that produces the desired results.

The modelers admit this in their articles published in peer reviewed journals. Stott, for example, in his 2000 *Science* article, says, "there are considerable uncertainties in some of the forcings used in this analysis," and "the good agreement between model simulation and observations could be due in part to a cancellation of errors." Hansen even went so far in his 1981 article to admit that the value for solar radiation used in the paper was chosen because, "other hypothesized solar radiation variations that we examined degrade the fit." It is not much of a trick to obtain a desired result when one can pick the values for the parameters that go into calculating it.

Of course, such frank admissions are never included when the media hypes the results of these types of studies. For example, press coverage of the Stott study in 2000 said it confirmed a hotter earth and that, "recent global warming is man-made and will continue."[30]

As a final check on the accuracy of climate model predictions, one can look at how global temperature actually has changed over the last ten years. Certainly, a prediction that global warming will continue is in keeping with the predictions of Figure SPM.5 in the IPCC's 2007 Summary for Policymakers[31] where all three scenarios show that global temperatures should steadily warm. Inspection of the GISS temperature record[32] (Figure 21) shows that instead of a steady warming, the fluctuating line tracing average global temperature has leveled off. Compared to 1998, the temperature in 2008 was almost 0.2 degree C (0.36 degree F) cooler. At the time this is written, as carbon dioxide increases, the cooling trend has continued into 2009 as

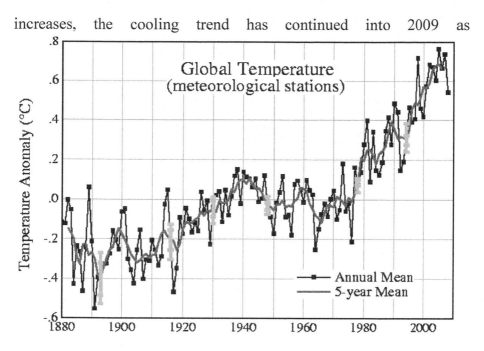

Figure 21. GISS global temperature, 1880-2008.

satellite data of the lower troposphere temperature clearly shows (Figure 22.)[33]

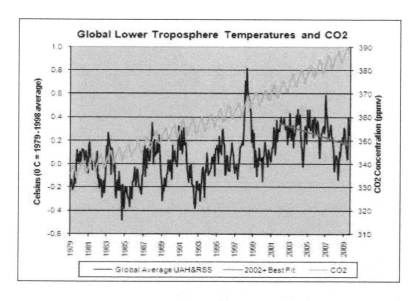

Figure 22. Lower troposphere temperature and CO_2

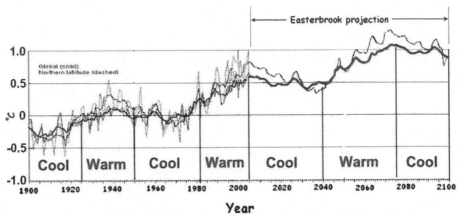

Although climate models did not predict the current global temperature decrease, professor Don Easterbrook did. As a geologist at Western Washington University, he knows how much climates have changed in the past and he is impressed with the many correlations between temperature and various solar cycles. Unlike GCMs, his graph of future climate[34] is based on past climate patterns. It shows the current cooling phase and predicts it will last until about 2040. His model predicts only about 0.4 degrees C warming by 2100, far less than the IPCC.

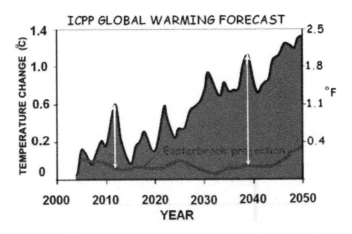

Figure 23. Easterbrook and IPCC projections of temperature

Climate change always zigzags—you said so yourself—so we can't expect climate models to predict wiggles, just the general trend. And a few cooler years does not change the fact that the trend is still up, exactly like GCMs project.

Take another look at the NASA graph in the previous chapter showing global temperature since 1880. It is obvious that the so-called warming trend since then is actually composed of wiggles, with none of them being longer than about 30 years. It is also obvious that the latest trend is toward cooling. No standard climate model predicted it, nor did the IPCC. It is opposite to the trend of carbon dioxide. For the United States, it was the 34th coldest summer on record.[35] The cold was centered in the north central states.

It's hard to admit that climate models might not be as reliable as the IPCC and the media say. Consider the recent comments of physicist, mathematician and Renaissance Man Freeman Dyson. "I have studied their climate models and know what they can do. The models solve the equations of fluid dynamics and do a very good job of describing the fluid motions of the atmosphere and the oceans. They do a very poor job of describing the clouds, the dust, the chemistry and the biology of fields, farms and forests. They do not begin to describe the real world that we live in."[36]

SUMMARY

Climate modelers regularly churn out what they refer to as projections of future global climate using general circulation models (GCMs) so complex that the world's largest supercomputer are required to run them. Although the modelers object that their projections are only scenarios, not predictions, a recent survey shows that nearly a third of climate model professionals equate the term "projection" with "probable."

The IPCC features these forecasts in their publications, especially highlighting them in their slick reports intended for government leaders and policymakers, and they lie at the heart of all their gloomy pronouncements. As a result, GCMs have become the best known and most powerful type of evidence used to support the global warming theory. They are based on scientists' understanding of the involved physical processes, and as is true of any computer simulation, their output is only as good as that understanding. Although the IPCC and the media treats their predictions as reliable, scientists' understanding of the chaotic controls of the earth's climate is limited and the uncertainty great.

The great physical uncertainty lying at the heart of GCM predictions is not part of the widely publicized IPCC graphs illustrating their predictions of future climate. Instead captions accompanying the graphs refer to standard deviation ranges of model averages, and this range of error (+/- 68%) is indicated on the graphs as shaded regions. Because the shaded areas make little difference in the amount of predicted warming over the next century, the average person gets the intended idea, that the predictions are based on sound science and are accurate. Such captions, however, refer to the precision of individual model runs, not physical accuracy. Run the program 100 times, and 68 times the prediction will be within the shaded region.

Dozens of physical parameters go into the construction of a general circulation model. Each and every one of them has some sort of error or uncertainty, and these are cumulative. For many of the physical processes that the climate models depend on, the uncertainty is far greater than is generally realized, as much as fifteen times the supposed amount of climate forcing attributed to human use of coal and petroleum. It is unscientific to claim "scenarios" showing future warming have any validity at all when the errors involved in the procedure are so much greater than the effect shown on the graphs. It smacks of IPCC advocacy or worse to misleadingly label such graphs with experimental precision while cloaking the high degree of physical uncertainty in obscure supplemental material.

Studies have been published which attempt to determine how well GCMs predict temperature and other aspects of climate. Results of such studies generally show climate models faring poorly. According to the most advanced GCMs in 1988, the earth should have warmed steadily since then, already being a degree or two C warmer now than is actually the case. Climate models also failed to predict accurately actual temperature trends in the troposphere, measured temperatures in Siberia over the past 125 years, in mainland Antarctica over 35 years and how doubling carbon dioxide in the atmosphere might effect the amplitude and frequency of El Niños. Of course, GCMs also did not foresee the current fluctuation toward cooling, although a geological model did. Unlike GCMs, this geologic model takes into account the relationship between past temperature trends and solar variability One wonders if general circulation models have ever successfully predicted anything.

Modelers say yes and cite the successful reproduction of climate over the past one hundred years, but only if the supposed human influences on climate is included. It sounds like a powerful argument because it seems so straight forward and reasonable, and it would be if the physical uncertainties weren't so huge, but that is not reality. Because the variables that are used in programming climate models are so great, a modeler can choose from a range of values for carbon dioxide forcing until one produces the desired results. Some skeptics say this amounts to nothing more than curve fitting, calculating a curve that best fits an existing series of data points.

It's a wonder the global warming theory ever got off the ground when its strongest supporting evidence, its gem, is actually so weak.

NOTES AND SOURCES

(1) For more information on the survey of climate scientists, see: http://scx.sagepub.com/cgi/content/abstract/30/4/534 .

(2) http://www.thefreedictionary.com/projection

(3) Skeptic, Vol. 14, p. 31, May 2008, Tapio Schneider, "How we know global warming is real."

(4) IPCC Third Assessment Report, p. 771, 2001, "The scientific basis contribution of Working Group I to the Second Assessment Report of the Intergovernmental Panel on Climate Change, Cambridge University Press.

(5) Nature weblog, 2007, K. Trenberth, "Predictions of climate," http://blogs.nature.com/climatefeedback/2007/06/predictions_of_climate.html.(6) The IPCC's 2007 Summary for Policymakers is available as a .pdf file at the following site: http://ipcc-wg1.ucar.edu/wg1/Report/AR4WG1_Print_SPM.pdf Figure SPM.5 is shown on page 14 of the report. The eighteen-page report can be downloaded.

(7) A good discussion of standard deviation is here: http://en.wikipedia.org/wiki/Standard_deviation(8) Skeptic, Vol. 14, p. 22, May 2008, Patrick Frank, "A climate of belief."(9) Ibid

(10) Climate Research, Vol. 18, p. 259, 2001, W. S. Soon et al, "Modeling climatic effects of anthropogenic carbon dioxide emissions: unknowns and uncertainties."

(11) The IPCC Fourth Assessment Report, 2007, Supplemental Materials for Chapter 8, Figures S8.5, S8.7 and S8.14: http://www.ipcc.ch/pdf/assessment-report/ar4/wg1/ar4-wg1-chapter8-supp-material.pdf CAUTION - this is a 41 MB download, but it is worth the wait. Save it to your hard drive for future reference.

(12) Geophysical Research Letters, Vol. 25, p. 2715, 1998, G. E. Myhre et al, "New estimates of radiative forcing due to well mixed greenhouse gases."

(13) The chart Dr. Hansen used in his speech to Congress in 1988 is reproduced at this site: http://www.climate-skeptic.com/2008/06/great-moments-in.html. A link at this site is provided to Hansen's text and chart.

(14) Dead Heat, the Race Against Global Warming, 268 pages, 1990, Basic Books, Inc. Publishers, New York, Michael Oppenheimer & Robert Boyle.

(15) Climate Research, Vol. 25, p. 185, 2004, T.N. Chase et al, "Likelihood of rapidly increasing surface temperature unaccompanied by strong warming in the free troposphere."

(16) Ibid, #6

(17) Geophysical Research Letters, Vol. 29, p. 1029, 24 Sept. 2002, I.V. Polyakov et al, "Observationally based assessment of polar amplification of global warming."

(18) Nature, Vol. 415, p. 517, 2002, P.T. Doran et al, "Antarctic climate cooling and terrestrial ecosystem response."

(19) Journal of Climate, Vol 19, p. 694, 2006, W.J. Merryfield, "Changes to ENSO under carbon dioxide doubling in a multimodel ensemble."

(20) Ibid, #8

(21) Ibid, #12

(22) *Global and Planetary Change,* Vol. 37, p. 103, 2003, C. Covey et al, "An overview of results from the Coupled Model Intercomparison Project."

(23) Figure 2a along with the rest of Grank's article is available on the Skeptic website: http://www.skeptic.com/the magazine/featured articles/v14n01 climate of belief.html

(24) *Climate Dynamics,* Vol. 83, p. 106, 2001, S.J. Lambert & G.J. Boer, "CMIP₁ evaluation and intercomparison of coupled models."

(25) *Ibid,* #23.

(26) *Ibid,* #23.

(27) *Discover,* June, p. 38, 2009, Corey S. Powell, "The big heat," quoting Bill Easterling, first column, p. 43.

(28) *Science,* Vol. 213, p. 957, 1981, J. Hansen et al, "Climate impact of increasing atmospheric carbon dioxide."

(29) *Science,* Vol. 290, p. 2133, 2000, P.A. Stott et al, "External control of 20th century temperature by natural and anthropogenic forcings."

(30) Quotation is from an article available at:
http://www.co2science.org/articles/V4/N4/EDIT.php

(31) *Ibid,* #6.

(32) The GISS average global temperature record through 2008 is available here:
http://data.giss.nasa.gov/gistemp/graphs/Fig.A.lrg.gif

(33) A graph showing satellite data of lower troposphere temperature is updated monthly at the Friends of Science web site. Click on the small graph for a larger version: http://www.friendsofscience.org/

(34) Professor Easterbook's article "Climate Change in the Coming Century," and the graph of his predicted temperature trend is available on his web site:
http://www.ac.wwu.edu/~dbunny/research/global/predictions.pdf

(35) See NOAA's announcement here:
http://www.noaanews.noaa.gov/stories2009/20090910_summerstats.html

(36) The quotation from Professor Freeman Dyson along with more of his comments about climate models and modeling in general is available here:
http://www.nationalpost.com/news/story.html?id=985641c9-8594-43c2-802d-947d65555e8e

CHAPTER SEVEN—CORRELATIONS AND GRAPHS

There are facts we must face. Carbon dioxide is the most important human-emitted greenhouse gas, and its atmospheric concentration continues to increase every year. The IPCC 2007 Report for Policymakers said it has increased from a pre-industrial value of about 280 ppm to 379 ppm in 2005. The same report says that this far exceeds the natural range over the last 650,000 years, and that its rate of increase is accelerating. Now it's up to about 388. The report says that with *very high confidence*, human activities have caused the earth to warm since 1750. As Al Gore says, aren't these "inconvenient truths?"

These statements mix facts with suppositions lacking observational backing. Sure, this is what the latest IPCC Report for Policymakers says.[1] And yes, just a glance at the graph of atmospheric carbon dioxide as measured in Hawaii[2] shows it is increasing (Figure 24).

Based on polar ice core proxies the statement about carbon dioxide in the last 650,000 years is also correct, although, the IPCC doesn't bother telling policymakers that carbon dioxide far exceeded current levels numerous times earlier than 650,000 years BP. The last statement, however, lacking any direct evidence, is pure advocacy. A more honest statement would be along the lines: human-emitted greenhouse gases may have contributed to the warming during the twentieth, but our knowledge of the chaotic and highly complex climate system is so poor, we cannot state that with any degree of scientific certainty beyond pure speculation. Without doubt, labeling it "very high confidence," 90% certain according to the IPCC, lacks any veracity at all.

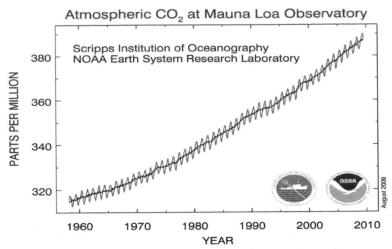

Figure 24. Carbon dioxide in the atmosphere, Mauna Loa record.

There is direct evidence that carbon dioxide and other anthropogenic greenhouse gases cause global warming. Everyone has seen it. Everyone is aware of it because its been in newspapers and magazines as well as on TV. Plot temperature and carbon dioxide on a graph over the past 130 years, and it's clear. There is a well established correlation between increasing carbon dioxide and increasing global temperature.

Oh, yes, a graph like Figure 25 from zFact.com.[3] Mean global temperature in degrees F since 1880 is plotted on one vertical axis and carbon dioxide, in parts per million, is shown, in red, along the other vertical axis. A quick glance is all it takes to notice the upward trend of carbon dioxide and temperature, the correlation that is so well known.

Correlation in casual conversation is used to mean co-relation, in other words, one variable in a relationship changes, and by so doing, causes the other variable to also change.[4] This is unfortunate because direct causality is not implied when scientific reports refer to correlation or the correlation coefficient. A correlation, as the term is used in science, refers to a relation between two variables such that as one of the variables changes, so does the other. The correlation coefficient is a calculated statistical function that assigns a mathematical value to the correlation. The correlation is positive if both variables change in the same direction and negative if one changes opposite to the other.

There are actually several types of correlations that can be

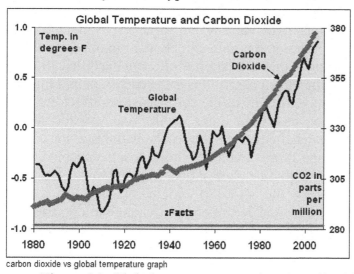

carbon dioxide vs global temperature graph

Figure 25. Global temperature and carbon dioxide

calculated, but the one most commonly encountered is the Pearson correlation coefficient. Calculation of this type of correlation coefficient always results in a number ranging from -1 to +1. A value of 1 is a perfect correlation and 0 means there is absolutely no relationship between the two variables, i.e., the change in one is completely independent of the other. A value of -1 also means perfection of correlation, but in the opposite direction. Squaring the correlation coefficient and multiplying the result by 100 expresses as a percent the amount of variability between the two variables that explained by the correlation coefficient. For example, squaring a correlation coefficient of 0.7, considered a strong one, and multiplying by 100 produces 49%. This means that 49% of the variability is explained, but 51% is not explained.

If the above two variables are plotted on a graph, their points will scatter about a straight line. With a correlation of 1, the points would exactly fall on the straight line, and there would be no relation to a straight line at all if the correlation is 0.

It is *always* a mistake to reach the conclusion that one variable directly *causes* the change in the second variable. The correlation coefficient only expresses the amount of variability between two variables, showing the distance of the points from a straight line when they are plotted on a graph. In the example above where 49% of the variability is accounted for, one of three possible conclusions has to be correct:

1. Change in the first variable causes the change in the second.

2. Change in the second variable causes the change in the first.

3. No causality is involved. The two variables, by chance, happen to change together, either in the same direction, or in opposite directions.

There is nothing in the correlation coefficient itself that allows one to decide which of the three possibilities is the correct one. It must also be remembered that even if one of the variables does affect the other, the effect may not be direct. In other words, one variable may effect something else which then effects the second variable in the correlation, causing it to change. Armed with only how close points on a graph plot with respect to a straight line, there is absolutely no way determine any of this.

In the real world of the earth's climate, change in one variable may produce a change in the second, which then may cause more change in the first or it may counteract the change in the first. The manifestation of this possibility in climate science is known as a feedback, either positive or negative. Feedback mechanisms, mentioned in Chapter Five, are just one of the many factors making it such a formidable task to understand climate change.

Maybe correlation doesn't automatically equate to cause, but aren't there situations where it can strongly suggest it? Say we plotted the age of growing children on a graph against their height. There would be some scatter, some points that seem out of place, but if we had a large number of data points, they'd fall pretty close to a straight line. Wouldn't we be justified in saying that growth as children age causes the increase in height?

In this overly simplistic example, that is certainly the case, but we know this already, from prior knowledge so the correlation is trivial. Without that prior knowledge, we would have no basis from the correlation by itself to draw such a conclusion. If we do so anyway, for whatever reason, we are making a mistake in logic. Consider a real world example.

Until 2002, doctors prescribed estrogen replacement in middle-age women as a therapy that lessened the symptoms of menopause. It was also thought that the therapy had an added benefit, that it protected menopausal women from an increased risk of coronary heart disease. This situation changed in 2002 when a controlled study involving a large number of middle-aged women was published that showed the hormone therapy actually increased the risk of heart disease.[5]

What caused such a mistake to be made? Previous investigations found that women taking the estrogen therapy had fewer heart attacks than women not on the therapy. In other words, taking estrogen correlated with less heart disease. Even though the researchers knew that correlation does not equate with cause, they thought in this case it did, since the cause seemed obvious because heart disease certainly did not cause the estrogen therapy. Restudying the data that had led to the erroneous conclusion showed that far more women on the hormone therapy came from a higher socio-economic class than the population as a whole. They exercised more and had better diets in comparison to the larger group. It was these which had produced the positive correlation, not the estrogen therapy, as was shown when these factors were controlled in the larger study.

This should clearly illustrate why, even when it seems obvious, correlation by itself never equates with cause. That is not to say that a correlation can never be used as a clue or hint to what the cause might be, a direction in which to conduct research. Cause, however, must always come from the research, not the correlation by itself.

Well of course, that's a sound principle, but there is another important principle, one that a great many people believe should be applied to serious environmental problems, the Precautionary Principle. It states that when a dangerous problem is identified and the consequences of doing nothing are serious enough, action should be taken to reduce the threat even if scientific cause and effect has not been fully established. Global warming is just such a threat. Temperature and carbon dioxide increase together on that graph, and we know carbon dioxide is a potent greenhouse gas. Maybe that correlation doesn't prove causality, but it surely gives us a strong hint; besides, what else could it be? If we wait around for proof of the obvious, it will be too late to prevent the catastrophe.

Perhaps examples could be cited where acting emotionally rather than rationally has prevented disasters, but on balance, the lesson of history is that rash action causes many more catastrophes than it prevents. However, instead of history, let's turn to science. Turn back two pages and take a closer look at the graph showing temperature and carbon dioxide.[6] This particular graph is referenced not because it is outstanding, but, on the contrary, because it is such a typical example of similar graphs that have been widely publicized in the mass media.

Temperature is shown in degrees F, not as the actual mean global temperature, but as a departure from zero, telling us the data is based on some climatic normal for temperature, although we are left to guess which normal period was used. The next thing to note is that carbon dioxide, shown in ppm, has steadily increased since 1880, not always at the same rate, but never a long reversal in trend. According to this graph, the rate started to accelerate in the late 1950s. This is probably related to opening of the Mauna Loa carbon dioxide observatory at that time, making real data available for the first time on atmospheric carbon dioxide. We are left to guess the source of the plotted values for carbon dioxide before that. Is it proxy data from ice cores, an informed estimate, or somebody's guess?

Note also the sharp increase in temperature beginning in the late 1970s and how it coincides with the steeper rate of carbon dioxide increase, highly indicative of cause and effect in most people's minds. The temperature line on this graph shows no sign of turning down, but then it conveniently ends several years ago before the current cooling became obvious. A clearer picture of what is really a rather poor correlation between temperature and carbon dioxide would be even more apparent than it is already if the graph were updated, but maybe clarity is not this web site's intention.

Notice the period from 1880 to about 1912 and the period from 1943 to the late 1970s. Temperature declines during both of those time frames even though carbon dioxide keeps increasing. In other words, for more than 60 years, about half of the time shown on the graph, the trend in temperature is opposite to the trend in carbon dioxide, highly suggestive that other factors, unidentified factors, cause the temperature trend. These other controls, whether they be natural or artificial, might simply sometimes overpower the effect of carbon dioxide, or the effect of carbon dioxide might be small or even nonexistent. The graph raises these possibilities, but does not provide any information that would allow us to answer the questions. Since this chapter focuses on correlations, a fuller discussion of the relationship between carbon dioxide and temperature is saved for Chapter Nine.

Concerning the correlation between carbon dioxide and temperature, the graph, as incomplete as it is, still makes it clear that the correlation is not nearly as strong as people have been led to believe. One variable whose trend runs in the opposite direction from another is not the stuff of a strong correlation.

A fair point, but it might be related to the lack of solid data on carbon dioxide before measurements started at Mauna Loa.

What if we focus on the steep rise in temperature along with carbon dioxide starting in the late 1970s?

Indeed, if there is a correlation, this is where it should be strongest. A Finish study investigated the time of sharpest increase in temperature, 1979 through 1998, a twenty-year period when the surface record shows a warming trend of 0.188 degrees C per year (0.338 degrees F)[7]. Although carbon dioxide increased each year (Figure 26), the annual rate varied considerably, from 0.70 ppmv (1993) to 2.88 ppmv.(1998). Subtracting the mean rate of increase over this twenty-year period from the annual rate of increase gave what the authors called the "Carbon Dioxide Thermometer." Treating surface temperature the same way, the calculated correlation coefficient between temperature and carbon dioxide is only 0.46. Squaring this and multiplying by 100 tells us that the correlation accounts for just over 20% of the variability, leaving nearly 80% as unexplained. A casual glance at the graph[8] might lead one to suspect a more significant correlation, but the zigzag pattern of temperature is the culprit.

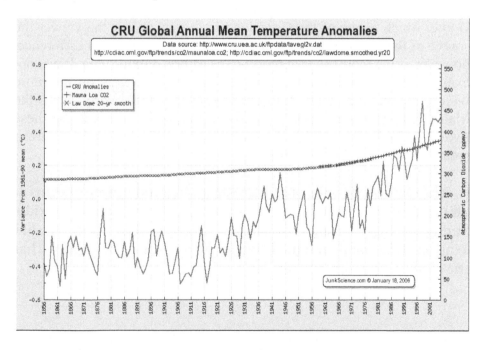

Figure 26, Temperature and carbon dioxide trends, 1979-1996

There's another important point that should be considered concerning the widely reproduced carbon dioxide-temperature graphs like the one we've been talking about. One must be wary of graphs. Although they seem straightforward, they can easily mislead, particularly with just a quick glance. The number one rule is to always be aware of the numbers that are used. Read the caption carefully and consider why a certain scale was chosen rather than another. Would a more accurate representation of the data result from using a different scale? Why does the data on the graph start and end where it does? Was the graph constructed to fairly represent the data or to support some bias? For instance, take a look at another graph of global surface temperature plotted against atmospheric carbon dioxide.[9]

At first glance, there doesn't seem to be much of a correlation between temperature and carbon dioxide on this graph. It this because different data is plotted?

No, they're both showing virtually the same information. This one goes back to 1856 for temperature, but other than that, the data is the same. It also is a far superior graph because it shows the viewer where the Mauna Loa data starts being plotted on the graph (marked by where the plus signs on the carbon dioxide line begins, i.e., the darker portion of the line). It also gives the source for the earlier carbon dioxide data, Law Dome, which is one of the Greenland ice cores, telling the viewer the source of the proxy data that's plotted. The really important difference, however, is the scale used for plotting the carbon dioxide data. It was not exaggerated to make it look like a gigantic increase, as on the more widely reproduced graph. Without that, the increase early on in the Mauna Loa data does not look nearly so impressive. Comparing these two graphs makes a good illustration of how simply varying the scale can create graphs that produce completely different impressions.

The scale used for carbon dioxide on the first graph[10] begins at 280 ppm, widely cited as the preindustrial level of carbon dioxide, and goes high enough to show current levels. How is that exaggerating the problem? It could just as easily be argued that the second graph minimizes it.

There is a subtle difference between the two that can easily be missed, something that is a widely used trick in "lying with graphs." The scale used for carbon dioxide on the popular graph does not begin at zero. This technique is widely used when the person constructing the graph wants to create the impression in the viewer's mind that something has

started increasing at a tremendous rate compared to the past. By leaving out the zero, it's easy to create the impression of a much sharper increase than actually exists. Introductory lessons on graphs in school teach people to watch out for this, but it is still a widely used method to create a false impression, and not just with global warming. For instance, consider the example shown in the graph above concerning housing prices.[11]

One might think that the IPCC would be above using such an old trick, considering how the media treats the IPCC as authorative and neutral, but they are not, especially in their less technical publications. A good example is Figure SPM.1 on page 3 of their latest Summary for Policymakers.[12] Three colorful and eye-catching graphs, reproduced on the top of p. 110 (reproduced as Figure 27) show the change in concentrations of three greenhouse gases over the last 10,000 years, carbon dioxide, methane and nitrous oxide. Each appears to have increased explosively in the twentieth century and each graph lacks a zero on the vertical scale showing gas concentration.

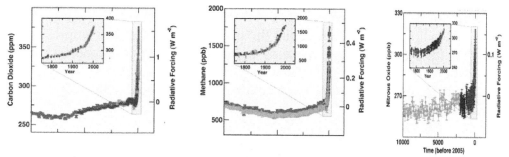

Figure 27. IPCC graph from Summary for Policymakers.

A similar graph showing the same gases over the past 2,000 years, also without a zero, is found on p. 100 of the Frequently Asked Questions section of the IPCC's latest report (Reproduced as Figure 28,).[13]

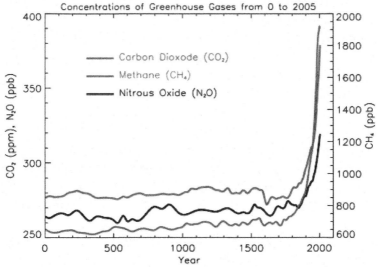

Figure 28. Projected increase in three gases, from IPCC

Another interesting and subtle example of how to promote a false impression using a graph can be found in the IPCC's 2007 Summary for Policy- makers, Figure SPM.3 on page six.[14] (Figure 29) The graph plots "Global Average Temperature" since 1850 in relation to average temperature for the period 1961-1990, the zero value. The problem here is not the graph itself, rather it is the interpretation in the text on page five: "The linear warming trend over the last 50 years...is nearly twice that for the last 100 years."

Reading that, and inspecting the graph, creates the clear impression that the curve showing warming is becoming steeper, supporting the notion that the rate of global warming is accelerating. One wonders how the IPCC determined this, but no explanation is offered in the Report for Policymakers itself.

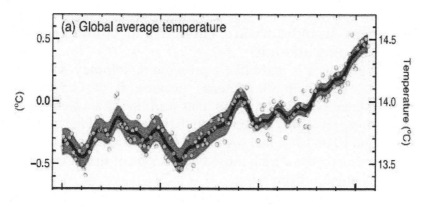

Figure 29. IPCC global average temperature.

The answer is found in FAQ 3.1, Figure 1 on page 104 in the Frequently Asked Questions supplement to the IPCC's 2007 report. The graph showing temperature since 1850 from the SPM is repeated, but now a series of straight lines is drawn on the various sections (Figure 30). The caption explains the straight lines: "Annual global mean-observed temperatures (black dots) along with simple fits to the data." In other words, they have drawn straight lines through selected sections of the graph, the last 25, 50, 100 and 150-year periods. The caption continues, "Note that for shorter recent periods, the slope is greater, indicating accelerated warming."

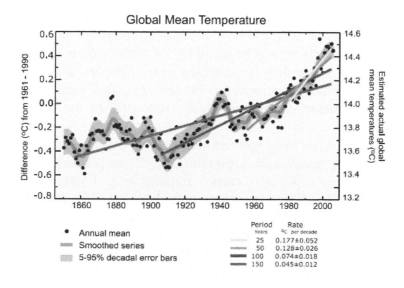

Figure 30. From IPCC FAQ supplement

That seems straightforward enough and completely valid. What's misleading about it?

Misleading? No, it is more like a prosecuting attorney withholding evidence that will help the defense because it creates the false impression, stated plainly in the text, that warming is accelerating. It is equally valid to draw straight lines through the section of the graph from 1858-1878 and 1910-1943, and say that the steepest rates of warming for the past 150 years were during those two periods of time, so therefore, the rate of warming is decreasing.

Scientists striving for impartiality would never attempt such a blatant misrepresentation, but for politicians trying to change public policy, it is business as usual.

SUMMARY

In everyday conversation, the term "correlation" is related to cause. In science, however, correlation refers to a statistical function, the Pearson Correlation Coefficient, a measure of the linear relationship between two variables, such as carbon dioxide and global temperature. The correlation is said to be positive if both variables change in the same direction and negative if one changes opposite to the other.

Calculating the Pearson Correlation Coefficient, the type most commonly encountered, always produces a value ranging from -1 to +1. A value of 1 is a perfect correlation and 0 means there is absolutely no relationship between the two variables. Squaring the correlation coefficient and multiplying the result by 100 expresses, as a percent, the amount of the variability between the two variables. For instance, a correlation coefficient between two variables of 0.7, generally considered a highly significant correlation, shows that 49% of the variability is explained.

Although it tends to be ignored in the global warming controversy in relation to temperature and carbon dioxide, the well known saying that "correlation does not imply cause" remains true. Establishing that a positive correlation between increasing carbon dioxide in the atmosphere and global temperature insures that one of the three following possibilities is correct: 1) carbon dioxide causes the temperature increase; 2) higher temperature causes the carbon dioxide increase; 3) neither of these two variables is directly related to the other. Given only that the correlation is positive does not allow any choice to be made as to which of the three is correct. However, the correlation can be treated as a clue or hint as one searches for additional physical evidence.

Close inspection of popular graphs showing carbon dioxide and temperature, often cited as proof that humans are causing the earth to warm dangerously, offer no direct evidence of that at all. Close inspection of such graphs shows the correlation between the two is actually poor. For about half the time interval typically charted, temperature actually decreased while carbon dioxide continued to increase. The closest correspondence between the two variables over the last 150 years is from the late 1970s to the early years of the twenty-first century, but even for this time period, the correlation is poor, less than 0.5.

One must be careful in evaluating graphs constructed to illustrate temperature and greenhouse gas relationships to avoid false impressions. It is particularly important to always read the explanation and to be aware of the numerical numbers being graphed. Intentional deception is always possible so do not assume the particular scale used for the graph was chosen to most accurately represent the data. One of the most common ways to mislead a casual viewer is to not start one axis at zero. This makes it easy to give the impression that the variable being graphed has suddenly started to increase at an explosive rate. Examples are cited in IPCC reports of this exact fault as well as more subtle ways to deceive.

NOTES AND SOURCES

(1) The IPCC's 2007 Summary for Policymakers is available as a .pdf file at the following site: http://ipcc-wg1.ucar.edu/wg1/Report/AR4WG1_Print_SPM.pdf The eighteen-page report can be downloaded.
(2) The Mauna Loa Observatory's graph of carbon dioxide is available: here: http://www.esrl.noaa.gov/gmd/ccgg/trends/co2_data_mlo.html
(3) A typical example of a graph showing carbon dioxide and average global temperature is available at this site: http://zfacts.com/p/226.html
(4) See the article on correlation at Wikipedia: http://en.wikipedia.org/wiki/Correlation
(5) Mayo Clinic, "Hormone therapy: Is it right for you?" Available at: http://www.mayoclinic.com/health/hormone-therapy/WO00046
(6) *Ibid*, #3
(7) The text and data tables for the Finish study is posted here: http://www.john-daly.com/co2-conc/co2therm.htm
(8) Ibid, #3.
(9) The graph is available at the Junk Science web site: http://www.junkscience.com/Greenhouse/cause.html
(10) *Ibid*, #3

(11) Several examples of misleading graphs are discussed at this website: http://www.coolschool.ca/lor/AMA11/unit1/U01L02.htm
(12) The IPCC's 2007 Summary for Policymakers is available as a .pdf file at the following site: http://ipcc-wg1.ucar.edu/wg1/Report/AR4WG1_Print_SPM.pdf Figure SPM.1 is on page 3.
(13) Available on the web at: IPCC Report/AR4WG1 FAQs.pdf
(14) *Ibid*, #12

CHAPTER EIGHT—THERMOMETERS, SATELLITES, THE SUN AND COSMIC RAYS

How is the global average temperature actually determined?

The existence of reliable temperature data is often taken for granted in the global warming controversy. It's a subject that receives very little media attention because it seems almost trivial in the age of computers, cell phones and space craft. Thermometers have been around for a long time, and they can be constructed to register temperature very accurately, so why should any problem exist? Yet, even something that seems as simple as this turns out to be quite controversial.

The controversy exists because determining the average temperature of the earth is actually a Herculean task that is fraught with many potential uncertainties and problems. Since we are land-dwelling creatures, we tend to think of global temperature as referring to the continents where weather stations exist, forgetting that over 70% of the earth is ocean. This means that the air temperature over water is going to count more than twice as much as the land in deriving average global temperature, and how in the world can we determine that? Well, we can't really, so citing one number for global temperature is really a fiction.

Obviously, the problems that exist with land-based data will be different than those for the oceans. Because the amount of temperature change is small, measured in tenths of degrees, any systematic negative errors could completely mask warming. If the errors were positive, warming would be suspected when none actually occurred.

The difficulties are magnified because we are not just interested in current temperatures, but in a historical record. Weather stations were never randomly located, a requirement for obtaining scientific data. Instead, they tended to be placed near where people lived, introducing the problem of poor or uneven coverage. Fewer weather stations existed in the nineteenth and early twentieth centuries than today. Many of the early stations, originally sited in rural areas, now find themselves in towns, suburban areas or even surrounded by cities where the urban heat island effect might artificially warm the data.

Even when the stations have remained rural, the instruments themselves may have changed as well as the protocol for taking a reading. For instance, recording the high and low temperature for one twenty-four-hour period will produce different results from always reading the temperature at a certain time or times of the day. How the data is obtained has also varied, opening the shelter door to read the thermometer compared to modern automatic readings. The possibility exists that such variations might affect the recorded temperatures. Physical degradation of the shelter itself might also affect the data, not to mention any lack of care in obtaining or recording the data.

Similar problems affect the ocean temperature record. In addition, the problem of lacking uniform geographic coverage is magnified because shipping lanes have been anything but random. To overcome such problems, satellites were launched in the 1970s to record various types of physical data including temperature. Far more uniform coverage of the globe resulted, but instead of ending the debate about global temperatures, satellites brought with them new controversies. Because they measure temperature in the lower part of the troposphere rather than the surface, one issue is how one record relates to the other. After satellite data over several years became available showing less warming than surface records, climate alarmists questioned the reliability of the satellite data. This controversy still simmers.

Although both sides in the global warming controversy agree that such problems can seriously distort temperature data, they do no agree on how to calculate and apply adjustment algorithms to remove whatever bias might exist in the data. The theory underlying such procedures is sound if the bias is fully understood and accurately measured, but of course, this in itself becomes another area of controversy. Global warming skeptics charge that the generally applied adjustments underestimate the bias in the data resulting in an over-estimation of warming. A recent peer-reviewed article concluded that half of the global temperature trend over land in the late twentieth century is due to such underestimation.[1] This conclusion is supported by a peer-reviewed study of proxy tree-ring density during the twentieth century,[2] which found considerably less warming than the generally accepted instrumental record.

Is some official agency, perhaps one connected to the United Nations ,such as the IPCC, responsible for maintaining global temperature records?

There are government agencies, but no organization has responsibility for one agreed-upon worldwide data set. There is, in fact no single data set that is universally recognized the world over as be "the one" to use. Perhaps the World Meteorological Organization comes closest to fulfilling such a role. It is the organization of the United Nations that supplies information to its 188 member nations, whose governments are mostly global-warming-alarmist. The WMO's main activities include hosting conferences and promoting cooperation among nations in exchanging meteorological data. In this role, it helped establish the IPCC, but the WMO functions more as a coordinator of world climate data rather than an originator of data.

Worldwide, five "official" sets of thermometer data are maintained: GISTEMP, NCDC, HadCRUT3, RSS MSU and UAH MSU. The first three of these are ground based records, while the final two are remote sensing data records for the troposphere and stratosphere. All are in the United States except HadCRUT3 which is located in the United Kingdom. A handy comparison of the recent data from three of them is posted at JunkScience.com (Figure 31).[3]

GISTEMP, NASA's Goddard Institute of Space Studies

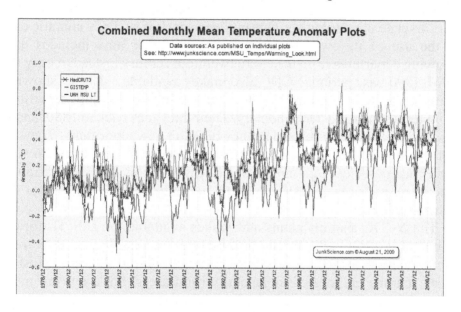

Figure 31. JunkScience.com's comparison of GISTEMP, NCDC, HadCRUT3 temperature records.

temperature record, which begins in 1880, is probably the best known and most widely sited. Closely connected with James Hansen, of the five records, its global temperatures have tended to be the warmest, which makes it a favorite of global warming alarmists while skeptics hint at intentional bias. For years, the media made a big splash over GISTEMP's proclamation that 1998 was the hottest year in the entire record until 2007, when mathematician Steve McIntyre, publisher of the Climate Audit blog, found a mistake in how the temperatures were calculated.

After NASA corrected the data, most of the hottest years turned out to be in the 1930s, not the 1990s. Of course, the elite media did not consider this worthy of publicizing. In 2008, GISTEMP showed October as especially hot, but had to retract this pronouncement when Steve McIntyre discovered they had substituted September data for October.[4] Critics attributed both of these errors to bias toward a warming climate, just as they do to NASA's constant revisions to the data sets, even from years far in the past, in ways that increase the current amount of warming.

NCDC, the National Climatic Data Center, is a branch of NOAA. Its two main divisions are the World Data Center for Meteorology, in Asheville, NC, known as the "world's largest active archive of weather data," and the World Data Center for Paleoclimatology in Boulder, CO. NCDC works closely with several regional climate centers and the various state climatologists. One unique aspect of NCDC's climatic data is the use of its own correction algorithms that now includes data showing departures from the entire twentieth century instead of only the 1961-1990 base period. NCDC also makes available a data set showing actual mean temperatures, NCDC Absolute, instead of only departures from the average, NCDC Anomaly. Their data goes back to 1880, and is also being constantly revised as they develop new algorithms. They, of course, say their procedures improve the data, but some question why each change seems to go only in one direction, showing more warming. The result is that the NCDC is pulling out in front of GISTEMP for the warmest record.

The NCDC also maintains and makes available the U.S. Historical Climatology Network of daily and monthly meteorological data from more than a thousand stations across the contiguous 48 states. Most of these stations are located in generally rural areas, but a few are more urban. The length of the period of record varies, but many extend back into the nineteenth century. The data is available in raw form and after various adjustments have been performed. An NCDC web page describes the rationale for the adjustment procedures.[5] The adjustments have the effect of increasing the temperature, as shown clearly in a brief video.[6] Convenient access to the data set is available on the web.[7]

The UK's Met Office, Hadley Center for Climate Change, maintains the HadCRUT3 gridded data set of near-surface temperatures Going back thirty years earlier than the others, to 1850, it is based on a global network of long-term land stations and regular measurements from ships and buoys. Using their own set of correction procedures, some global warming skeptics consider the product to be a more honest representation of global temperature.

The two remaining records are Microwave Sounding Unit (MSU) satellite records of tropospheric and stratospheric temperature extending back to late 1978. There have been a series of satellites providing the data over the years, each having its own calibration problems due to such factors as slight differences in sensors and orbital decay. Part of the controversy concerning the satellite record revolves around the techniques used to knit the data together to make a continuous record in order to spot trends. Most of the controversy centers on one of the three main channels of data, MSU 2LT, a record of the lower troposphere temperature.

Two principle groups are involved with the effort, RSS, Remote Sensing Systems, a research oriented company located in Santa Rosa, CA, that is principally funded by NASA's Earth Science Enterprise program, and the UAH group, the University of Alabama at Huntsville. Each group makes its own decisions as to the appropriate procedures for stitching the data together, resulting in slightly varying data, another area of controversy. Global warming skeptics gravitate toward the UAH dataset because it tends to be somewhat cooler. Climate alarmists prefer RSS data suspecting bias in the UAH data because the group is headed by Roy Spencer and John Christy, two well-known and active global warming skeptics.

As mentioned, much of the controversy concerning the satellite record over the years has centered on the UAH MSU 2LT channel because it found temperatures in the lower portion of the troposphere to be lower than the RSS record and lower than climate models predicted. The trend from this record showed much less warming, or even cooling, compared to surface temperature records. Not surprisingly, climate alarmists argued that the RSS record was the correct one because it better fit climate models.

The MSU 2LT channel of data is itself more open to criticism because it is based on techniques for subtracting the stratospheric influence (MSU-4) from the MSU-2 record, middle troposphere, to produce a lower troposphere record. There was quite a heated exchange of research papers for a few years, some claiming errors in the UAH data processing, other claiming the suggested corrections themselves were in error. The UAH group did find one error in their data related to satellite orbital decay. Correcting it decreased the differences between the two groups, but the minor change did little to resolve the controversy. From May 2008-May 2009, RSS found the lower troposphere temperature averaging

0.09 degrees C above the long term mean while UAH for the same period showed +0.05 degrees C. Such small differences lead to big controversies in the global warming debate.

The UAH satellite record[8] shows a slight upward trend in lower lower troposphere superimposed on the usual zigzag pattern of oscillating temperature. By far the warmest year is 1998. A small cooling trend started in this record in 2002 and continues currently at a slightly increased pace.

Comparing this record to the one from RSS,[9] plotted at the same scale, shown in Figure 32, makes one wonder why so much has been written comparing the two datasets because the pattern is very similar. Again, 1998 shows a sharp peak and a cooling trend since 2002 is apparent. Both records show less warming than the surface datasets, which was the underlying reason for the controversy over the satellite data. Of course, the surface temperature record also shows less warming than predicted, but this wasn't true a few years ago, before we entered into the present cooling phase.

Global warming alarmists argued that because the satellite data didn't show sufficient warming, something had to be wrong with the satellite data. In other words, turn science on its head by elevating theory to being a more important indicator of accuracy than actual data, a strange argument that reverses the high priority science accords data. Nevertheless, a paper published in the peer-reviewed journal *Science* in 2003 used this logic to choose the RSS dataset over the UAH[10].

The methodology of the article compared the two satellite records through 2001 with the output of a sophisticated Department of Energy general circulation model, the PCM model. An accompanying graph from the article[11] (shown as Figure 33) compares the two satellite data sets to the PCM predicted trend in the stratosphere (Graph A) and troposphere (Graph B). Inspection of the graph shows that the two satellite datasets are remarkably close to each other, corresponding almost exactly for most of the record for both stratospheric and tropospheric temperatures. In contrast, the PCM simulation has only a slight resemblance to the actual data, particularly in the troposphere where for 1983, 1987-'88 and 1998, the predicted trends are opposite to the real trends. Nevertheless, after some sophisticated statistical calculations, the

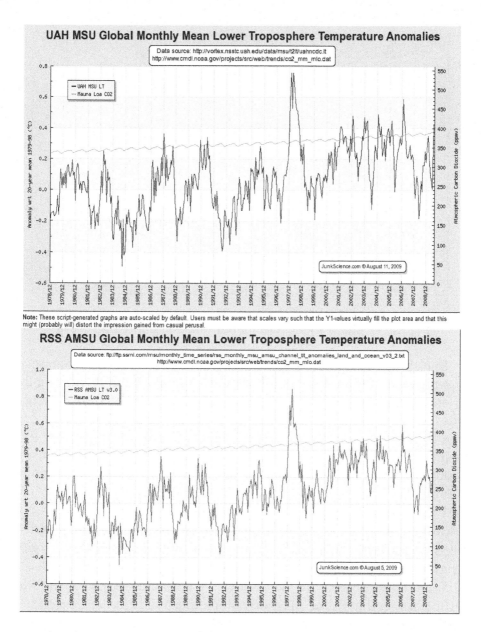

Figure 32. Comparison of UAH and RSS temperature records.

authors argue that their computer-generated PCM model, shows the RSS data is closer to reality, and of course, that it indicates human-caused warming.

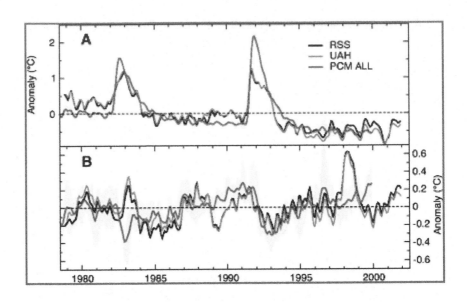

Figure 33. PCM model compared to UAH and RSS records

Back when Freon was the great global menace and the ozone scare was big news, the so-called ozone hole was supposed to be causing the stratosphere to cool. This was a plausible explanation if ozone really was decreasing in the stratosphere because ozone is a greenhouse gas. The current idea is that an increased greenhouse effects traps more heat near the surface, leaving less to reach the high stratosphere, causing cooling. This, of course, depends on the stratosphere actually undergoing cooling.

I've read that the upper atmosphere, the stratosphere, is cooling while the troposphere warms just as climate models predict, and that this is dramatic proof of climate models as well as greenhouse warming.

Well, let's see what the data shows. The Hadley center of the Met Office has an interesting diagram comparing trends in the troposphere and stratosphere (reproduced as Figure 34).[12] Satellite data, both RSS and UAH, goes back to 1978, and the data is extended back into the 1950s using balloon-carried radiosonde data, which agrees fairly well with the later satellite data. For comparison, surface data in green is also shown. As can clearly be seen, the lower troposphere has generally warmed during the sixty years of this record. General circulation models do predict warming of the troposphere, in fact, they predict a good bit more warming than has actually occurred.

Compare the surface record in the lower graph (shown in green) with the stratospheric record in the upper graph. Note that in the ground record, most of the hottest years are after the eruption of Mt. Pinatubo in 1992. If the global warming alarmists are right, this is when the greenhouse effect should be really high, resulting in a maximum of stratospheric cooling, but a glance at the upper diagram shows this is not the case. There was a cooling trend in the stratosphere for half a century, but over the last ten years, there's been no cooling in the upper atmosphere.

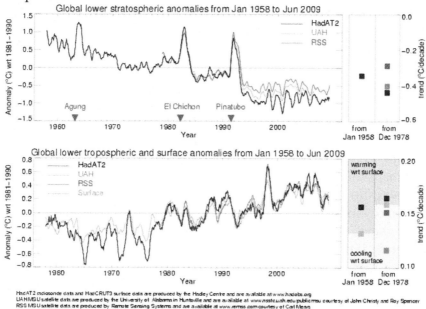

Figure 34. Tropospheric and stratospheric temperature anomalies

An obvious explanation would be recovery of the ozone layer from governmental banning of Freon, but alarmists can't say that because they don't want to let up on attacking chlorine/bromine compounds.[13]

Global warming skeptics say that the amount of surface warming since the nineteenth century is partly due to urbanization. Is there anything to that argument?

While skeptics and alarmists agree that what's called the urban heat island effect is real, they disagree about its importance. Alarmists say it's minor, not very important at all, but skeptics say it can account for much of the warming. It would seem that research should be able to solve this question, but both sides of the debate can point to results supporting their position.

Anyone who lives in a large city knows how much cooler the air feels, winter or summer, out in the rural countryside. This is due to removing the natural vegetation cover in heavily populated areas, paving large areas with heat-trapping asphalt and concrete and adding large numbers of heat sources-- buildings, houses, air conditioner vents, generators and such. All of these together constitute the urban heat island effect.

Certainly there has been a lot of population growth since the nineteenth century with widespread expansion of urban areas out into the surrounding land. Rural areas have become suburban and suburbs have become urban in many areas of the world, but has the effect been enough to affect the overall global temperature? A study in the peer-reviewed *Journal of Climate* found that it has not[14], but another study in another peer-reviewed journal found that it was a significant effect, enough to account for 2/3 of the accepted amount of global warming since the nineteenth century.[15] These two serve as examples of the many more studies that each side in this debate can cite to bolster their position. If nothing else, the literature on the subject makes it clear that when it comes to urban heat islands, the science is far from settled.

Perhaps there is no way to settle this issue, but there does exist one very interesting long-term temperature record that speaks to it, Armagh Observatory in Ireland. The observatory is on top of a 200-foot hill in an area of woods and parkland that has changed little since the observatory was founded in 1790. There has been little population growth in the area and with a wind often blowing off the nearby sea, any urbanization effects on the temperature record should be minimized. This makes it virtually unique in the entire world, a reliable long-term instrument record of temperature free of contamination due to an urban heat island effect.

Inspection of the record[16] (Figure 35) shows very little warming, with most of it being due to an increase in the low temperature, rather than an increase in the high. In other words, the little bit of warming that has occurred in this area of northern Ireland is due to milder nights, but warmer days. Comparing this record to another long term record, the Hadley Center's Central England Temperature (CET) Record[17] is interesting. The area the CET record covers, England's industrial heartland, is likely to have been affected by increasing urban heat, so corrections have been applied. Even so, it is obvious that considerably more warming has taken place in the CET record than Armagh. This suggests that urbanization may be having a greater effect on the global surface temperature than is generally accepted.

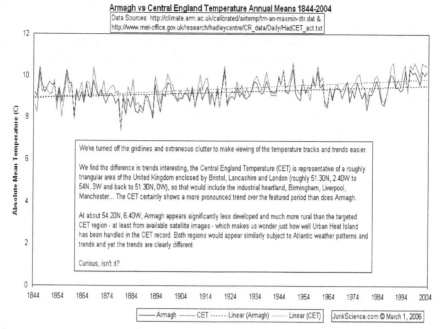

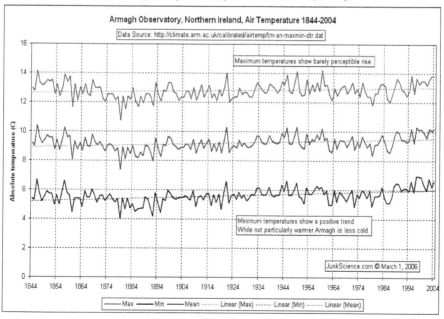

Figure 35. Comparison of Armagh and CET temperature records.

Both these records only have local significance because they don't look anything like one for the earth as a whole that goes back a thousand years. It's in Al Gore's movie and book, and it shows the globe is a lot warmer now than anytime over the past millennium.

Ah, yes, the famous graph that looks like a hockey stick (Figure 36). Stephen Schneider kept it on his Stanford University web site[18] until he died in 2010, and it was featured in the IPCC's 2001 Summary for Policymakers. Most of the diagram is based on a synthesis of various types of proxy data for the northern hemisphere such as tree rings, pollen analysis and coral. It shows no sign of the Medieval Warm Period a thousand years ago, or the Little Age. In fact, it shows little variation in temperature at all until the twentieth century when it turns up sharply like a hockey stick. The part that really makes the temperature trend look like a hockey stick are the huge projections of high temperature for the current century. The graph is based on the work of Michael Mann and others as published in 1998[19] and 1999.[20] Despite literally hundreds of peer-reviewed articles showing that temperatures in many parts of the world were warmer a thousand years ago than today, Mann's work showed no warming a thousand years ago, or at least, it was cooler at that time than the late twentieth century.

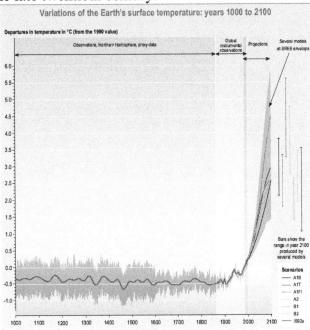

Figure 36. One version of the famous "hockey stick" diagram.

Typically in science, paradigm-changing research that overturns generally accepted knowledge is greeted with skepticism. Extraordinary standards of evidence are required and others have to be able to repeat the results, using the same methods. That didn't happen with Mann's work. The climate alarmist community greeted it with open arms because it did away with that pesky Medieval Warm Period. Alarmists never had a good answer for that and tried to ignore it. They admitted carbon dioxide was much lower at that time. If it was, how could the climate have been warmer? The obvious answer was that something else was at work, and if so, why couldn't it be at work now? Yes, life was going to be much easier without that warm period. *Scientific American* even listed Michael Mann as one of the top fifty visionaries in science in 2002.

It wasn't long however, before critics began to find flaws in the complex statistical technique Mann used to link together various kinds of proxy temperature from all over the northern hemisphere in order to form one single record. Foremost among them were mathematician Steve McIntyre and economist Ross McKitrick who claimed in a peer-reviewed journal article[21] that the statistical technique and computer program Mann used artificially created the hockey stick shape. McIntyre also found that Mann's study relied heavily on a series of bristlecone pine tree rings that were wrongly dated and interpreted. His critique eventually forced Mann to admit mistakes in print.[22]

The current status of the hockey stick graph depends on one's viewpoint. Global warming skeptics claim that it has been thoroughly discredited, so much so, that even the IPCC no longer uses it. Climate alarmists say that's not so. Their take is that despite the best efforts of critics, the best they could do was to find small errors that hardly affected the results, and that it doesn't matter much anyway because now others have achieved similar results using larger datasets.

The history of the hockey stick controversy makes for fascinating reading, certainly deserving of more than the brief summary given here. The story is easy to find on the web, but one must be careful. Climate alarmists sites, such as RealClimate, which was actually started by Mann and some colleagues, to defend their work and computer modeling in general, slant their articles in favor of the hockey stick characterizing it as being 100% valid, while global warming skeptic sites, for example Icecap, give the reader exactly the opposite spin, with the hockey stick being fatally flawed and 100% discredited. The two fairest and most complete accounts I'm aware of are listed in the references.[23]

As a geologist, what is your take on the hockey stick?

I tend a be a bit suspicious of research with results dependant on statistical manipulations so sophisticated ordinary people can't begin to follow how those results were derived. It seems to me that if there is a real result, or at least a significant result, such a procedure shouldn't be necessary.

In the case of Mann's original graph and subsequent work producing similar results, I think there's a more important reason to be cautious. The existence of the Little Ice Age and the warm climate a thousand years are about as well established as anything can be concerning the history of the earth, such as the existence of wooly mammoths during the Pleistocene and dinosaurs in the Jurassic. There are literally hundreds and hundreds of studies from all over the world, including land and sea, many of them made using geologic materials, that show the medieval warming. Such studies are so numerous, that the CO_2 Science web site provides a new link to one of these studies each week,[24] and they've been doing this for quite awhile. Yes, this is a skeptics web site, but all the linked studies are in peer-reviewed science journals. A good suggestion would be to check some of them and make up one's own mind. Beyond the many studies, there is also additional evidence of the medieval warming that anyone can see if they spend some time in the Arctic tundra. Too cold for trees now, there are numerous old tree stumps and branches. If not a thousand years ago, then when was it warm enough for these trees to have grown?

Beyond the large numbers of paleoclimatology studies, history itself attests to both these dramatic periods of climate. Records attest to the growth of European glaciers during the Little Ice Age. The advance of some even destroying villages that had existed in valleys for hundreds of years. There are many records of the poor harvests and ensuing widespread famines, and not just in Europe. Documents in New England show that snow fell on some places every month of the summer ruining the harvest. That was 1816, the famous year without a summer.

The medieval warming is equally well known. It is beyond question that a thousand years ago, it was warm enough in Greenland for the Vikings to establish three colonies. Archaeological research shows they fared well for a time until the climate started to slowly cool. Eventually sea ice around Greenland became so thick that supply ships could no longer come and go. Crops also failed and the colonists slowly perished.

Global warming alarmists discount all this sort of data saying these unusual climates were nothing more than local effects, not hemispheric or global. What they seem to forget is that global trends are nothing more than a large number of local trends. They also underestimate how widespread both these climate periods were.

There's a long-established convention in science known as Occam's Razor that seems to help in choosing between equally well-supported explanations. According to it, often the simpler of two equally supported explanations is the correct one, but of course, this is not absolute. Concerning the Medieval Warm Period, the choice is between the results of hundreds of well documented studies spanning many years of work conducted by a large number of scientists neutral in the global warming debate and a few new studies processing existing data using statistics and computer programs undertaken by scientists not impartial in the climate debate. Following Occam's Razor, I continue to accept the established results until solid evidence comes along convincing me they are all wrong.

Skeptics also say that possible effects of the sun on climate are ignored in global warming theory, but climate researchers say that of course, they consider the sun, but the effects are negligible. Which side is right?

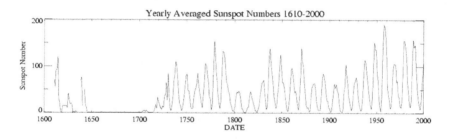

Figure 37, Yearly averaged sunspot numbers, 1610-2000, NASA

There are different aspects of solar energy production that must be considered. Of course, there is the well established link between the eleven-year sunspot cycle and climate mentioned in Chapter Three. Sunspots are dark areas of lower temperature on the surface of the sun, tending to be in certain latitudinal belts rather randomly distributed. Tiny in comparison to the sun's disk, even small ones are larger than the earth. Galileo was one of the first to see them, and a record began to be made, at first spotty, but improving quickly. As can be seen in this nice NASA chart (Figure 37)[25] the eleven-year cycle can vary by a couple of years, and the number of sunspots also varies, from very few or even zero, to nearly two hundred. Total solar irradiance is greater during times of many sunspots, as discussed in Chapter Three.

The most notable feature of the chart is the long time interval from 1645 to 1715 when there were almost no sunspots, the Maunder

Minimum, named for astronomer Edward W. Maunder who first noted it. There have been other minimums, such as the Dalton from 1790 to 1820, but none were as deep as the Maunder. This is true even when including the extended sunspot record made using a radioactive carbon (C^{14}) proxy.[26] Something strange was happening with the sun during the seventy years of the Maunder, or so the observational record indicates. Some scientists make the case that the observational record was still spotty then, making the anomaly less deep than it seems.

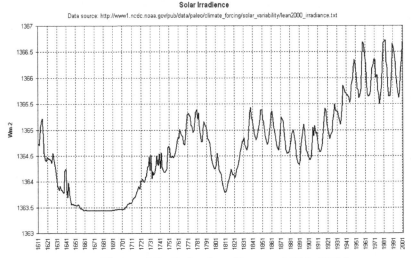

Figure 38. Solar irradiance, 1611-2001.

Although it might seem logical that solar irradiance would be less when there are many sunspots, the opposite is true because solar activity increases with sunspot number. In other words, when there are many of the cooler, dark splotches, larger areas of the sun grow hotter. As can clearly be seen in Figure 38,[27] solar irradiance dipped below 1363.5 W/m² during the Maunder Minimum. It's highest level since 1611 occurred during the late twentieth century with a TSI of about 1366.7 W/m². That's a difference of 3.2 W/m². In terms of climate change, the question is whether 3.2 W/m² is enough to affect climate.

It seems pretty small, way less than one percent, maybe a quarter of a percent, probably not enough to affect climate. Correct?

Once again, it depends on point of view. Global warming alarmists say the effect is negligible and not large enough to be important. Skeptics say it might be significant.

Looking again at the extended record of sunspots,[28] one can see that a pretty fair correlation exists with temperature. The Medieval Warm Period was a time of high solar activity, the Little Ice Age, with three sunspot minimums including the Maunder, was low in activity and the warming since the nineteenth century corresponds nicely to the modern maximum. Of course, just as is true of carbon dioxide, a positive correlation does not imply cause, but it might be a clue.

Perhaps the sun itself will provide the answer soon because it's showing signs of decreasing activity. The beginning of the next sunspot cycle, Cycle 24, is long overdue, and, as shown in Figure 39, the number of sunspots is unusually low. Some solar physicists think we may be about to enter a new solar minimum and are excited by the prospect. Among them is NASA scientist Dean Pesnell who says, "For the first time in history, we're getting to observe a deep solar minimum." Writing in the April 21, 2009 *Connecticut Examiner*, meteorologist Joe Roy is not so sure. "As of April 19th, there have no sunspots observed on 96 out of the 109 days (88%) so far this year. However, sunspot activity has been lower on several occasions during the last 400 years. There is a better chance than not that sunspot activity increases in the next few months.[29]" If it doesn't however, he goes on to say, "According to NASA research, there is a cause-and-effect relationship between sunspot activity and measured changes in the global temperatures on earth. I strongly agree."

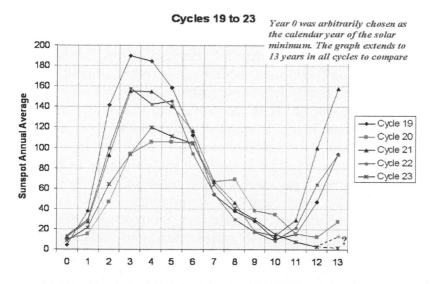

Figure 39. Comparison of recent sunspot cycles

As this is being written, the sun seems headed toward its deepest sunspot minimum since the nineteenth century. Comparing the end stages of the previous four sunspot cycles to Cycle 23 shows the unusually long wait for 24 to begin.[30] Cycle 23 has already exceeded 12.5 years. The last time a cycle this long occurred was 1848. One has to go back to the Dalton Minimum in 1816, the so-called year without a summer, for a longer cycle, 12.7 years. NOTE - See ADDENDUM, BACK OF BOOK.

If we are at the beginning of a new sunspot minimum, how will this affect temperature?

No one is certain, but the answer should be interesting. Global warming alarmists say the effect will be minimal, perhaps enough to counteract greenhouse warming for a few years, but then warming will become really vicious. Skeptics say the correlation between global warming and solar irradiance is better than with carbon dioxide and that we may very well be at the beginning of a long cooling spell. They point out that the 3.2 W/m^2 difference between the Maunder Minimum and the active sun of the late twentieth century is greater than both the generally accepted 2.7 W/m^2 increase in the greenhouse effect that the IPCC attributes to human emissions since 1900 and their estimated 1.6 W/m^2 total positive forcing (see Chapter Six).

As will be discussed in the next chapter, global warming alarmists argue that positive feedbacks considerably magnify this 2.7W/m^2 increase in the greenhouse effect causing the earth to warm. Surprising enough, a Danish scientist has proposed a new theory in which cosmic rays serve as a positive feedback to increase the effect of variations in solar irradiance.

The theory comes from Henrik Svensmark of the Danish National Space Center. As Svensmark explains in a recent summary article,[31] cosmic rays, high energy subatomic particles from other stars, such as He nucleons, cause ionization in the lower portion of the troposphere, releasing electrons. These electrons serve as condensation nuclei, which are necessary for the formation of clouds. In other words, a greater the intensity of cosmic rays causes more reflective clouds to form in the lower troposphere. With more clouds, more of the sun's energy is reflected back into space instead of warming the ground, so the globe cools. When there are fewer cosmic rays, more of the solar irradiance reaches the surface and the earth warms. Svensmark and co-workers first proposed this process as a theoretical construct in 1995, but it has now been experimentally verified.

What is perhaps most intriguing about this new theory is how variations in solar irradiance produces a positive feedback with the cosmic ray flux. A more active sun, that is, many sunspots, produces increased shielding for the earth, decreasing the cosmic ray flux, so that more of the sun's energy reaches the surface, amplifying the warming effect. During times of fewer sunspots, such as the Maunder Minimum, the reverse occurs, less shielding and more clouds, so the cooling is deepened.

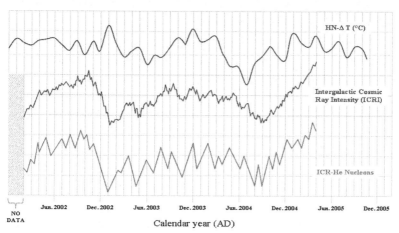

Data Source: Intergalactic Cosmic Ray discovered by Voyager 1, credit NASA. ΔT by NOAA/NASA/JPL/GSFC. Stone et all. *Voyager 1 Explores the Termination Shock Region and the Heliosheath Beyond.* Science, Vol. 309, Issue 5743, 2017-2020, ©23 September 2005. Graphs interpreted by Nasif Nahle © 2005.

Figure 40. Cosmic ray intensity and temperature.

A connection between cosmic rays from deep space and the earth's climate is a completely new idea, totally unexpected. Today, the theory's primary evidence is experimental results and graphs (Figure 40) showing that the cosmic ray intensity closely tracks temperature.[32]

It should be no surprise that climate alarmists have not welcomed Svensmark's ideas with open arms. A theory that exploding stars in remote corners of the galaxy cause climate change, not our own excesses? Why, of course that can't be right. Their main argument is that such a positive feedback process has yet to be demonstrated under natural conditions. They are right about that, but that's also true when it comes to the positive feedback process they propose between carbon dioxide and water vapor, as will be discussed in the next chapter.

Both of these positive feedback processes are theoretical, based on known physical quantities. Neither has been conclusively demonstrated to actually be important as a control of our climate under natural conditions. Until this has been completed and generally accepted, neither should be considered to be part of established science.

SUMMARY

Although reliable data concerning the average global surface temperature seems to be taken for granted in the global warming controversy, determining it is actually a Herculean task. Few weather stations existed before the late nineteenth century and they were never randomly located as needed to obtain statistically reliable data. Many were located on the edges of population centers and were later engulfed as cities expanded, raising the possibility of biasing the data due to artificial warming. Even the few that have remained rural have their own problems such as changes in the instruments or manner of recording the readings. Problems are even worse for the oceans which constitute more than 70% of the earth's surface area. Data for the past oceanic temperatures is sparse, confined to a few scientific expeditions and scattered records from the shipping lanes. Because the amount of warming since 1880 is just a few tenths of a degree, it is certainly possible that some of it could be due to poor data.

Worldwide, five "official" global temperature data sets are maintained, three based on ground stations and two relying on satellite data: GISTEMP, NCDC, HadCRUT3, RSS MSU and UAH MSU. GISTEMP, NASA's Goddard Institute of Space Studies' surface based record has shown the warmest temperatures, but is now falling behind NCDC, the National Climatic Data Center, a branch of NOAA because of new adjustment procedures. The UK's Met Office, Hadley Center for Climate Change, maintains the HadCRUT3 data set of near-surface temperatures, extending back to 1850, thirty years more than the others.

Satellite coverage of tropospheric and stratospheric temperature started late in 1978. Two different groups, RSS MSU, Remote Sensing Systems, and UAH MSU, University of Alabama at Huntsville, work with the satellite data. MSU in the names refers to the Microwave Sending Unit. Although satellite coverage is much closer to being global than is the surface-based data, climate alarmists made it controversial because it showed less warming, or even cooling, compared to surface data. They made much of the fact that the UAH temperatures for the lower troposphere were lower than the RSS data because of differences in how the two groups process the data. Although true, the differences are tiny, at most a few hundredths of a degree.

Ironically, global warming alarmists use the satellite data to claim it supports greenhouse warming because the stratosphere is cooling but the lower troposphere is warming, just as it should in an enhanced greenhouse world. This sounds good, but stratospheric cooling stopped ten years ago, and even if it had not, there is another more likely explanation.

Global warming skeptics find the adjustment of surface temperature data controversial. Each of the three ground based data sets continually undergoes adjustment to compensate for urban warming and other effects. Skeptics say the adjustments underestimate "heat island" warming and they question why the corrections seem to go in only direction, toward showing more warming. The data from Armagh Observatory in Ireland, which has remained rural for more than 200 years, shows only slight warming, with a tendency toward warmer nights, rather than hotter days.

The famed "hockey stick" temperature graph lacking a Medieval Warm Period and Little Ice Age and showing no warming for the past thousand years until the twentieth century is especially controversial. According to global warming skeptics, the statistical and programming procedures used to stitch together several types of proxy temperature data from several areas included mistakes and questionable data. Alarmists respond that the mistakes do not change the conclusion and that more recent studies have produced similar results. Accepting the conclusion requires throwing out the results of hundreds of studies showing these distinct climatic periods did occur. It would also require ignoring historical records, as well as the observation that today's Arctic tundra has numerous old tree stumps and branches.

Climate alarmists say they do include the effects solar variations in general circulations models, but the effects are too small to amount to much. Yet, skeptics point to a number of recent studies with improved instrumentation suggesting that variations in the sun's total irradiance is enough to account for a major portion of the warming since the end of the Little Ice Age. An observational record of the eleven-year sunspot cycle goes back to the early 1600s, and has been extended back another few hundred years using a radioactive carbon proxy. A plot of this data shows it correlates better with global temperature than does carbon dioxide. Times of cooler climate are associated with sunspot minimal, such as the Maunder Minimum in the late seventeenth and early eighteenth centuries.

Global warming alarmists discount this data saying the sunspot cycle does not produce nearly enough variation in solar energy to account for such changes in temperature. In this regard, Henrik Svensmark of the Danish National Space Center has introduced a controversial new theory involving cosmic rays from space that partially control the number of clouds in the lower troposphere. According to the theory, a less active sun, that is, few sunspots, produces decreased cosmic ray shielding for the earth, increasing the cosmic ray flux, so that more clouds form. The clouds reflect more of the sun's energy into space thereby amplifying the cooling effect from lower solar irradiance. Svensmark's theory, therefore, provides positive feedback. It is too early to know how this theory will fare, but it should not be dismissed out of hand as alarmists do. Positive feedback from water vapor, which they do think is important, is also unproven.

NOTES AND SOURCES

(1) *Journal of Geophysical Research*, Vol. 112, p. 1029, 2007, R.R. McKitrick & P.J. Michaels, "Quantifying the influence of anthro-pogenic surface processes and inhomogeneities on global data."
(2) *Quaternary Science Reviews*, Vol. 19, p. 87, 2000, K.R. Briffa, "Annual climate variability in the Holocene: interpreting the message of ancient trees.
(3) JunkScience.com's "Global Warming" at a glance is here: http://www.junkscience.com/MSU_Temps/Warming_Look.html
(4) http://www.debatepolitics.com/Environment/39781-hansens-giss-temp-data-wrong-again.html
(5) Background and adjustments to the U.S. Historical Climate Network is available here: http://cdiac.ornl.gov/epubs/ndp/ushcn/ndp019.html#tempdata
(6) A video showing how adjustments in temperature affect the U.S. Historical Climate Network: http://www.youtube.com/watch?v=nxDFAdkOs6Y
(7) Check various stations that are part of the U.S. Historical Climate Network here: http://www.co2science.org/data/ushcn/ushcn.php
(8) The UAH MSU satellite record of lower tropospheric temperature is here: http://www.junkscience.com/MSU_Temps/UAHMSUglobe.html
(9) RSS AMSU satellite record of lower tropospheric temperature: : http://www.junkscience.com/MSU_Temps/RSSglobe.html
(10) *Science*, Vol 300, no. 5623, p. 1280, 2003, B.D. Santer et al, "Influence of satellite data uncertainties on the detection of externally forced climate change."
(11) Graph from Santer et al article comparing satellite data to a GCM prediction: http://www.sciencemag.org/cgi/content/full/300/5623/1280/FIG1
(12) troposphere and stratosphere temperature trends compared: http://hadobs.metoffice.com/hadat/images/update_images/global_upper_air.png
(13) JunkScience.com has a good discussion of the stratosphere-greenhouse question: http://www.junkscience.com/MSU_Temps/stratosphere.htm
(14) *Journal of Climate*, Vol. 16, p. 2941, 2003, T.C. Peterson, "Assessment of urban versus rural in situ surface temperature in the contiguous United States: no differences found."
(15) *Journal of Geophysical Research-Atmospheres*, Dec. 2007, R.R. McKitrick & P.J. Michaels, "Quantifying the influence of anthropogenic surface processes and inhomogeneities on gridded global climate data."
(16) Armagh Observatory temperature record available at this site: http://www.junkscience.com/MSU_Temps/Armaghan.html
(17) This link compares the Armagh and Central England Temperature records: http://www.junkscience.com/MSU_Temps/Armagh_vs_CET.html

(18) The infamous "hockey stick" diagram:
http://stephenschneider.stanford.edu/Climate/Climate_Science/earthsSurfaceTemp.ht
ml

(19) *Nature*, Vol. 392, p. 779, 1998, Michael E. Mann et al, "Global-scale temperature patterns and climate forcing over the past six centuries."

(20) *Journal of Geophysical Research*, Vol. 26, p. 759, 1999, Michael E. Mann et al, "Northern Hemisphere temperatures during the past millennium: inference, uncertainties, and limitations."

(21) *Geophysical Research Letters*, Vol. 32, doi:10.1029/2004GL021750. 2005, S. McIntyre and R. McKitrick, "Hockey sticks, principal components, and spurious significance."

(22) *Nature*, Vol   July 1, 2004 Mann confirms mistake

(23) Reasonably impartial accounts of the hockey stick controversy are found at the following web sites:
http://www.junkscience.com/jan05/breaking_the_hockey_stick.htmlhttp://
http://www.junkscience.com/jan05/lone_gaspe_cedar.html
www.newscientist.com/article/mg18925431.400-climate-the-great-hockey-stick-
debate.html?page=1; The first two sites are parts 1 and 2 of an article by Marcel Crok published in Canada's *Financial Post* on January 27 and 28, 2005. A lot of detail is given and the slant is a bit toward the skeptics' view. The third article is from the March 18, 2006 issue of *New Scientist*, a British publication fully accepting the alarmist view, but this article, by Fred Pearce, is only somewhat slanted in that direction.

(24) The Medieval Warm Period record of the week is a quarter of the way down the page. http://www.co2science.org/index.php

(25) NASA chart showing the sunspot cycle since the early seventeenth century:
http://solarscience.msfc.nasa.gov/images/ssn_yearly.jpg

(26) Carbon 14 extended sunspot record extending back nearly 1200 years:
http://en.wikipedia.org/wiki/File:Carbon14_with_activity_labels.svg

(27) Graph showing reconstruction of solar irradiance since 1610 here:
http://www.junkscience.com/Greenhouse/irradiance.gif

(28) *Ibid*, #26

(29) The article in the *Connecticut Examiner* is posted on the web:
http://www.examiner.com/x-6362-Connecticut-Weather-Examiner~y2009m4d19-
Sunspots-where-have-they-gone

(30) Sunspot cycles 19 through 23 compared in a graph as this location:
http://icecap.us/images/uploads/CyCLELENGTH.jpg

(31) *Astronomy & Geophysics*, Vol. 48, p. 1.18, 2007, H. Svensmark, "Cosmoclimatology, a new theory emerges."

(32) Graph showing cosmic ray intensity and temperature:
http://biocab.org/Anomaly_ICR_001_110305.jpg

CHAPTER NINE—CARBON DIOXIDE AND TEMPERATURE

The evidence is strong that carbon dioxide causes the global temperature to change. A graphs in Al Gore's book showing carbon dioxide and temperature in an Antarctic ice core is especially convincing. What other interpretation could there be?

Gore's book does include the mentioned graphs, but they are presented in a distorted manner designed to achieve a political purpose, to make the viewer reach the exact conclusion the question reflects, that carbon dioxide causes the global temperature to change.

There are several problems with the graphs he uses. He omits any scale for temperature making it impossible to place a numerical value on the temperature line. His graph for carbon dioxide has no zero and when combined with a low upper limit for carbon dioxide, it runs off the scale from the present, making it seem disaster is right around the corner. Finally, he separates temperature and carbon dioxide, using a separate graph for each, making it look like they change exactly in tandem, one tracking the other. Then, just in case the reader still didn't get his point, he states directly, "When there is more carbon dioxide in the atmosphere, the temperature increases because more heat from the Sun is trapped inside."[1]

Laura David, the producer of Gore's academy-award-winning movie *An Inconvenient Truth*, uses the same data in her book *The Down-to-earth Guide to Global Warming*. Published by Orchard Books and intended to educate (indoctrinate?) children at home and in school, it takes Gore's distortion to the next level by presenting false information in graphical form. Looking at the graph (Figure 41) which in the first edition is on page 18,[2] carbon dioxide, shown in red, changes first followed by temperature, in blue, pounding home the point of Gore's book and movie, that carbon dioxide changes drive global temperature change. The problem, however, is that the two trends on the graph are mislabeled. Carbon dioxide is actually the blue line, and temperature is the red. Whether this was intentional, as some global warming skeptics claim, or nothing more than an unintended error, does not change the fact that readers of the first edition of this book are given information opposite to the facts.

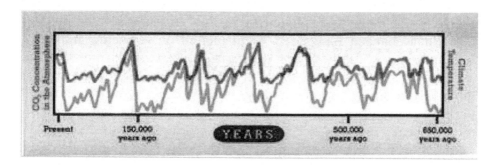

Figure 41. Diagram from Laura David's *Down to Earth Guide to Global Warming* with carbon dioxide and temperature reversed.

Are you saying in these ice core records, temperature changes before carbon dioxide? If so, why haven't I heard about it anywhere?

That's correct, temperature changes first. Typically several hundred years before carbon dioxide, but sometimes thousands. The same is also true of the other widely mentioned greenhouse gas, methane-- temperature changes first. This is no longer controversial in the scientific literature and is well known among climatologists. When the first ice core studies began to be published in the 1970s and 80s, the resolution was not sufficient to tell which changed first, but this gradually improved in later cores. That temperature changed before carbon dioxide began to become apparent when the initial Vostok 160,000-year-record was published in the 1990s (Figure 42).[3]

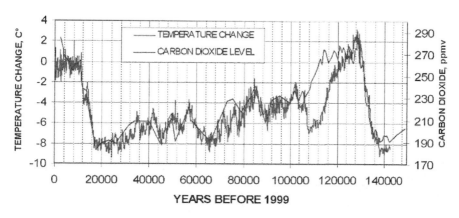

Figure 42. Vostok 160,000-year ice core record.

each other, so if they were plotted on separate graphs, it would be hard to tell which changed first. Even when plotted on the same graph, the relationship can be hard to spot, except in a few areas. The most obvious example in the Vostok graph is at 127,000 years BP where carbon dioxide starts up while temperature is going down. Instead of following, temperature continues down toward a trough at 108,000 years. Carbon dioxide finally reacts to this trend and heads down at 114,000 years. The relationship became increasingly clear with later studies, to the point where it could no longer be denied, although it can be hidden with carefully chosen graphs, such as this one from Wikipedia on Quaternary Glaciation,[4] (Figure 43) which strongly suggests that decreasing carbon dioxide caused the glacial advances

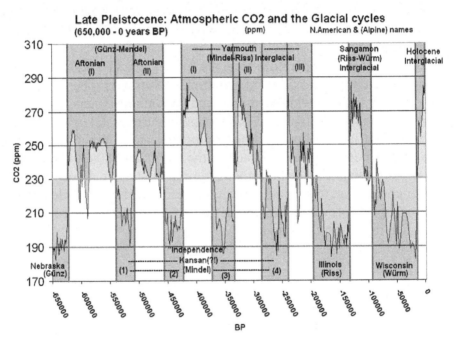

Figure 43. Wikipedia diagram - carbon dioxide and glacial cycles

and higher carbon dioxide led to the warm interglacials.

One of the first articles saying that the ice core records actually show temperature changing before carbon dioxide was published in 1988 in the peer-reviewed journal *Atmospheric Environment*.[5] Written by the well known global warming skeptic S.B. Idso, it was ignored. As better data became available, less controversial scientists began drawing attention to the same thing, including an article in the prestigious journal *Nature* in 1999.[6] These researchers concluded that, "the carbon dioxide decrease lags the temperature decrease," when glaciation is beginning and, "the same sequence of climate forcing operated during each termination." Several other articles along these same lines were published around this same time, but perhaps the definitive study is from 2003.[7] Using a far more precise dating method than previous studies, the authors concluded that "the increase in carbon dioxide lags Antarctic warming by 800 +/- 200 years."

It makes mo sense that temperature would change before carbon dioxide. We know carbon dioxide is an important greenhouse gas, so it must have some effect on global temperature. There must be some explanation for the contradiction.

That's exactly the reaction of climate alarmists when they could no longer refute the many studies showing temperature changed in the ice cores before carbon dioxide. Of course, what is supposed to happen in science is when a theory's key prediction is proven wrong, the theory is abandoned, but climate alarmists weren't about to let that happen. Instead, they chose the modification route, namely a positive feedback mechanism. The idea that emerged was that the Milankovitch orbital cycles nudged temperature in a certain direction. During the first few hundred years of this phase, temperature drives carbon dioxide, but then, the influence of carbon dioxide begins to assert itself through positive feedback so that it starts to force temperature. So for a few hundred years, temperature is in the driver's seat, but then for thousands of years, it's carbon dioxide.

The positive feedback mechanism that climate alarmists call upon to explain how carbon dioxide eventually drives temperature in the ice core data involves water vapor.[8] Recall from Chapter Two that water vapor is by far the most important greenhouse gas, although carbon dioxide greenhouse alarmists always tried to down play it. Luckily, it was still there when they needed it to patch up their tattered theory. This may be the only time they've actually acknowledged water vapor as such an important greenhouse gas. According to the proposed feedback mechanism, increasing global temperature, whether from carbon dioxide, orbital cycles, solar variation or whatever, will increase evaporation of water from the surface producing more water vapor in the atmosphere. More water vapor increases the greenhouse effect by adding to the major greenhouse gas. This causes more warming and more greenhouse gas in a self-sustaining positive feedback loop. RealClimate, a well known global warming alarmist web site, offers their own discussion of water vapor feedback.[9]

That seems like a reasonable process to me. It's well known that moisture evaporates more rapidly in warmer temperatures than cool. Why do you object?

Apart from the fact that such a self-sustaining process has not been demonstrated, once it got started, there seems to be nothing that could stop it. The earth should keep getting hotter and hotter until all the water in the oceans is evaporated. Eventually, it should be so hot that all the water vapor is lost to space and the earth became another Venus. Since nothing even remotely approaching this has happened in the entire 4.6-billion-year history of the earth, skepticism is warranted.

There's also easy-to-understand evidence, no math or physics required, against such a positive feedback process, the ice core data from Antarctica.[10]

Each peak in Figure 44 from Antarctica is a warm interglacial epoch,

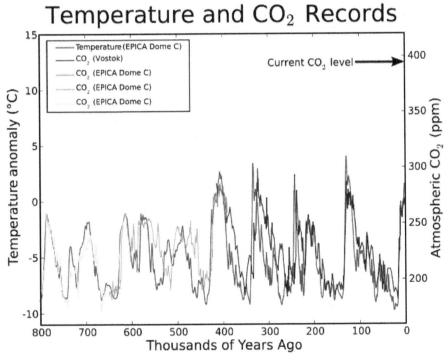

Figure 44. Carbon dioxide and glacial cycles, Antarctica.

and the valleys are the times of glacial advance. In each and every case, after thousands of years, the climate starts to warm. It reaches a peak, then turns down. Even at the compressed scale of this graph, several places can be seen where temperature changes before carbon dioxide, so if positive water vapor feedback is occurring, something else intervenes to bring the positive feedback to an end.

Another regular and easily observed temperature variation argues against water vapor actually being an important positive feedback control of other greenhouse gases. Take a look at Figure 45, a graph of satellite data showing average lower tropospheric temperature, 1979-1998,[11] in degrees Kelvin. The Kelvin scale is the same as Celsius except zero on the scale is -273 degrees C, absolute zero; therefore, 273 on the Kelvin is zero degrees C or 32 degrees F. Note that the graph does not show surface temperature.

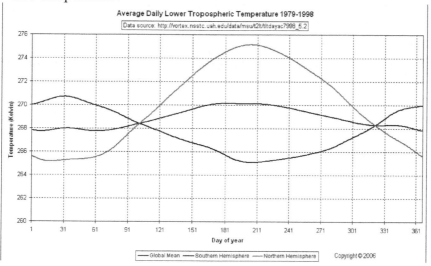

Figure 45. Lower tropospheric temperature, 1979-1998.

Seasonal tropospheric variation in temperature, the difference between winter and summer temperature caused by the earth's tilt on its axis with respect to the plane of its orbit around the sun, is obvious on the graph. Because land masses heat and cool much faster than the oceans and because the area of land in the northern hemisphere far exceeds that in the southern, the northern hemisphere dominates the yearly swing in global temperature. Notice that in the northern hemisphere, a nine-degree C (sixteen degree F) difference in temperature occurs each year. For the globe, the difference is smaller, but still more than four degrees F. Seasonal variation at the surface, as reported by the NCDC, is even greater, 6.8 degrees F.[12]

This is as reported by the <u>National Climatic Data Center</u>. the means 1961-1990 (commonly used as a reference period) work out the same globally but do differ slightly on a regional basis:

Combined Mean Surface Temp.	JAN	FEB	MAR	APR	MAY	JUN	JUL	AUG	SEP	OCT	NOV	DEC	Annual
1880 to 2004 (°C)	12.0	12.1	12.7	13.7	14.8	15.5	15.8	15.6	15.0	14.0	12.9	12.2	13.9
1880 to 2004 (°F)	53.6	53.9	54.9	56.7	58.6	59.9	60.4	60.1	59.0	57.1	55.2	54.0	57.0

So the earth already experiences a huge increase in temperature each year, much greater than the warming since 1880, and surely sufficient to trigger the positive feedback that increasing evaporation is supposed to have on greenhouse gases. Still, each year, global temperatures cool again, so if such a positive feedback mechanism is important, it's insufficient to overcome the combined effects of a changing angle of sunlight striking the surface and changing length of daylight, both resulting from the tilt of the axis of rotation.

Could there be some other explanation of why temperature changes before carbon dioxide in the ice cores?

One rather obvious explanation that isn't mentioned very often has to do with the carbon dioxide stored in the oceans. As discussed in Chapter Two, far more carbon dioxide is dissolved in ocean water than is present in the atmosphere   Carbon dioxide is actually a rare gas in the atmosphere, roughly equal to about 1 molecule for every 3,000 molecules of air. This is rather obvious if one stops to think about it for a moment because, as also discussed in Chapter Two, it displaces oxygen and becomes poisonous at higher concentrations, yet we inhale it with every breath, suffering no ill effects.

A common quality of liquids containing dissolved gases is that they can hold more gas at lower temperatures. A common example of this is effervescence of carbonated drinks, that is, the bubbles rising to the surface as the gas escapes from the liquid. Such drinks can hold less carbon dioxide when warm than when cold. The same is true of the carbon dioxide dissolved in the oceans. If they warm, for whatever reason, carbon dioxide will be released at a more rapid rate to the atmosphere. It would take some time, of course, for this to be detectable throughout the atmosphere, maybe several hundred years as the ice core records show.

Wouldn't this also be a positive feedback mechanism? The oceans warm, releasing carbon dioxide into the atmosphere, along with the carbon dioxide released from fossil fuels. Being a greenhouse gas, this increases the temperature even more.

Some global warming alarmists have made this argument, saying that as the oceans warm, even more carbon dioxide will get into the atmosphere, making it hotter than ever. For it to be valid, however, real world proof must be produced demonstrating that adding carbon dioxide to the atmosphere actually causes an increase in temperature. So far, the evidence shows the opposite. Take a look again at the initial Vostok ice core record.[13] Notice the peak warmth of the previous interglacial at just before 128,000 years BP. At that point, a sharp cooling trend begins that lasts for about 20,000 years, culminating in glaciation. But what happens with carbon dioxide? It oscillates up and down until finally, 16,000 years later, it drops sharply. During all that time, more than three times longer than all of human history, there is no relationship between the two. Next, look at the warming starting at 108,000 years until 102,000 years BP. Carbon dioxide drops throughout that time interval. Again, no relationship is demonstrated between carbon dioxide and temperature.

On the other hand, no place in the record shows increasing carbon dioxide followed by warming. There's no real world proof that carbon dioxide has been important as a control of temperature in this record.

One additional aspect of the ice core record deserves scrutiny. Alarmists make much of the fact that carbon dioxide was high during the interglacials and low during times of ice advance, attempting to imply that carbon dioxide is the control. However, they go on to overplay their hand by pointing out that carbon dioxide is a lot higher now than during the interglacials, so imagine how hot it's going to get. The graph[14] shows that the current temperature is about one degree C (1.8 degrees F) lower than in the previous interglacial. But what about carbon dioxide? It was about 290 ppmv 128,000 years ago, but now it's almost 390 ppmv. If carbon dioxide is such an important control of global temperature, alarmists need to explain how such a low level as 290 ppmv could have produced such a high temperature, and why it isn't much hotter now than it was then. Explaining this is even tougher if they try to bring in other greenhouse gases such as methane and nitrous oxide, which are also increasing. Alarmists would then have to explain how similarly low amounts of carbon dioxide are associated with the cold Little Ice Age and the very warm interglacial. How is that possible if it's the major control of global temperature?

Are you saying that it's all a big lie? That carbon dioxide doesn't have any effect on global temperature? That it's not a greenhouse gas?

Not at all. As discussed in Chapter Two, carbon dioxide is an important greenhouse gas, but it's not the most important one, and alarmists exaggerate its importance. To understand why, we have to get a bit more technical than we have so far. Recall that greenhouse gases absorb some of the sun's energy after it has heated the earth's surface and is being reradiated back toward space in the form of infrared energy.

Visible light and infrared are both of a continuous spectrum of energy known as the electromagnetic spectrum (Figure 46).[15] Visible light, located on the chart to the right of infrared, has a shorter wavelength.

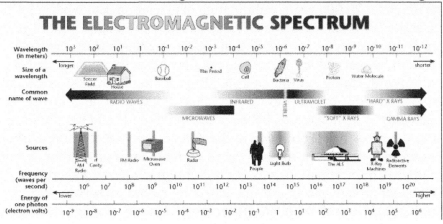

Figure 46. The electromagnetic spectrum.

A tremendously important but little understood point in the global warming controversy is that adding a lot more carbon dioxide to the atmosphere can only produce a small amount of warming.

The reason stems from a basic fact of physics. Carbon dioxide, as well as the other greenhouse gases, can only absorb infrared energy at certain wavelengths, not all wavelengths. Typically, several absorbing wavelengths (bands) occur for each greenhouse gas. The absorption bands of one greenhouse often overlap with those of another. This made relationship is made clearer by looking at Figure 47.[16]

It is important to understand what this chart shows. A scale is given for each gas, from 0 to 1. Zero means no absorption and a peak reaching 1 means complete absorption, 100%. No gas can absorb any more energy for any absorption band that has peaked at 1.

Notice first the total absorption for the atmosphere at the bottom and how most of that is accounted for by water vapor, by far the most important greenhouse gas. Compare the absorption bands of water vapor with carbon dioxide. There's a lot of overlap. Observe that all four of the absorption bands for carbon dioxide already overlap to various degrees with adsorption bands for water vapor.

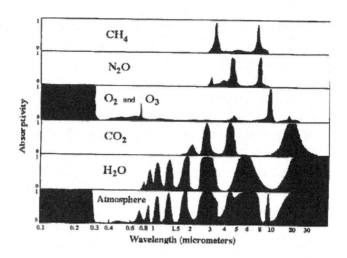

Figure 47. Absorption bands for various atmospheric gases

Understanding this is tremendously important for understanding the global warming debate. When a particular absorption band for any gas peaks at 1, *no more absorption can occur, no matter how much more of the gas might be added to the atmosphere.* For carbon dioxide, this means that adding more of the gas to the atmosphere cannot produce more absorption at three of its bands, no matter how high carbon dioxide might get, because they already peak at 1. Only the absorption band just above 2 micrometers is available for carbon dioxide, but it partially overlaps with water vapor; therefore, its absorption potential is limited because a sizable portion of the energy in this band is already being absorbed.

This can be illustrated with an analogy. Imagine walking along a white-sand beach on a bright sunny day. Putting on a pair of sunglasses helps to ease the glare. This works because the lens of the sunglasses absorbs a sizable portion of the light, so it's possible to see now without squinting. Imagine next putting on a second pair of sunglasses. The view becomes darker still, but the difference is not as great as from the first pair of sunglasses. Adding more pairs of sunglasses continues diminishing the light return until finally, no light passes through. With all the light now absorbed, adding another pair of sunglasses has no effect.

It is the same with any greenhouse gas. When an absorption band of a particular gas is saturated, adding more of that gas has zero effect. For carbon dioxide, this is the *real inconvenient truth*, the one that limits the effect of carbon dioxide as a greenhouse gas.

This seems reasonably clear. Carbon dioxide is an important greenhouse gas, but with limited available potential to cause

additional warming. If this is right, then the important question to answer is how much warming can carbon dioxide produce?

No precise answer to this question is possible, at our current state of knowledge, because there are too many unknown factors. The best that can be done is a reasonable estimate.

To do that, one additional fact needs to be understood. The effect adding more of a greenhouse gas to the atmosphere has on absorption is not linear, just like adding a second pair of sunglasses does not absorb as much light as adding the first pair did. In fact, the relationship is logarithmic. Wikipedia has a good introduction to logarithmic scales.[17] In practical terms, this means that adding the first portion of a greenhouse gas produces a large effect, but adding the same amount again produces an effect that is 10x weaker, and adding it a third times produces 100x less effect than the initial addition.

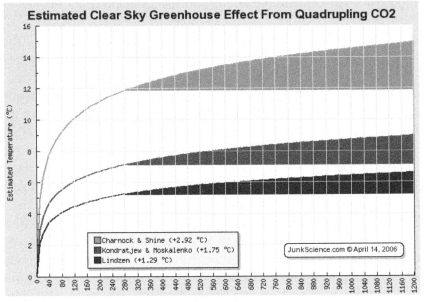

Figure 48. Estimated clear sky greenhouse effect from doubling CO^2

When facing such complex problems, scientists often try to simplify the situation by removing some of the complicating variables. Following such an approach, scientists have calculated the average global temperature if the atmosphere had no clouds and contained no carbon dioxide. Figure 48 uses this as the beginning point.[18.]

The diminishing effect of additional carbon dioxide is obvious because of the logarithmic relationship. Simply put, far more warming occurs for a given addition of carbon dioxide when the concentrations are low. Because the method assumes no clouds, it will *overestimate* the amount of warming more carbon dioxide can produce.[19]

With this caveat in mind, we can read from the graph[20] estimates of 0.6, 1.0 and 1.4 degrees C for doubling the current 388 ppmv of atmospheric carbon dioxide to 776 ppmv. Assuming water vapor accounts for 90% of the greenhouse effect (20% clouds and 70% absorption by water vapor) 10% is left for all the other greenhouse gases. Based on this, the lower estimate, 0.6 degrees C, is likely to be closer to the real answer.[21]

That's a lot lower than I've ever heard. Didn't the IPCC estimate it at something like 2 - 4.5 degrees C warming from doubling carbon dioxide? Why is their estimate so much higher?

Their estimates are based on the output of general circulation models discussed in Chapter Six. These computer models program in various types of positive feedback mechanisms which serve to greatly increase the amount of predicted warming, but particularly positive feedback resulting from increased water vapor. This is really where most of the warming in the GCMs come from, not carbon dioxide itself. As discussed earlier in this chapter, warming is supposed to increase evaporation resulting in more water vapor. This is reasonable as far as it goes, but it ignores one important fact. We are not sure whether increasing evaporation from the oceans would be a positive feedback or a negative.

A negative feedback? How could that be?

Remember a basic fact. Water vapor also forms clouds, as well as being a gas in the atmosphere. Clouds reflect a lot of the incoming solar energy right back into space before it ever has a chance to warm the surface, increasing the earth's albedo. Clouds also absorb a substantial portion of the solar energy that isn't reflected as it passes on down toward the earth's surface. These two effects account for cloudy days being cooler in the summer than clear days.

On the other hand, clouds increase the greenhouse effect tending to retain more of the energy that does reach the surface. This is why the temperature doesn't drop as much on cloudy nights in the winter as clear nights. Scientists still are not sure whether the net effect of more clouds is to warm or cool the earth, an ambiguity reflected in how GCMs deal with clouds. The possibility exists that more water vapor in the air is a negative feedback instead of a positive. If so, the main feedback used in GCMs is incorrect. The entire global warming theory would then crumble.

SUMMARY

The idea that carbon dioxide drives the earth's climate is firmly embedded in the public's collective mind. In support, reference is made to Antarctic ice core records, and popular books that say when there's more carbon dioxide in the atmosphere, temperature increases. The reality is the opposite. Temperature generally changes in detailed ice core records hundreds of years, or even thousands of years, before carbon dioxide (and methane). If there is a cause and effect relationship involved between carbon dioxide and temperature, temperature drives carbon dioxide, not the reverse.

Early ice core studies did not obtain sufficient resolution to tell whether temperature or carbon dioxide changed first in the records. Correlations between the two were generally quite good. Because carbon dioxide is a greenhouse gas, people worried about a warming planet naturally thought carbon dioxide drove temperature. Resolution improved sufficiently in the 1990s when the Vostok ice core from Antarctica was recovered showing there were problems with the earlier interpretations. With more and more detailed ice core records becoming available by 2003, it was shown that carbon dioxide usually lags behind temperature in the ice cores by several hundred years. This was well known when Al Gore's book was published in 2006, but it somehow escaped his attention.

The fact that temperature changes before carbon dioxide should have been sufficient to falsify the theory of global warming, but global warming is no ordinary scientific theory. As a *cause célèbre*, its many ardent supporters were not about to let it be vanquished so easily. Attempting to patch up the theory, they proposed that a positive feedback exists with water vapor. That's ironic since global warming skeptics had always tried to draw attention to water vapor as the most important greenhouse gas. The proposed mechanism is that in the ice core records, orbital changes nudge temperature in a certain direction, say warming. This causes more evaporation, releasing more water vapor and carbon dioxide into the atmosphere, which causes still more warming, and so on, a classic positive feedback process.

The central question is whether such a positive feedback process actually occurs. It has yet to be demonstrated for the earth's climate system, and it is possible that more water vapor in the atmosphere actually produces a negative feedback by increasing the number of clouds, which both reflect and absorb solar energy, causing cooling. Beyond this difficulty, there are other indications that such a process, if it occurs, is not important. Clear evidence from ice core records and the annual seasonal change in temperature shows that water vapor feedback does not lead to run away temperatures.

A central fact of physics is that greenhouse gases such as carbon dioxide *only* absorb infrared energy at certain wavelengths or bands. The effect is logarithmic. At first, adding a little carbon dioxide greatly increases absorption, but as concentration of the gas increases, much greater amounts of carbon dioxide are required to produce the same amount of absorption, until the maximum absorption potential is reached. After that, adding more carbon dioxide has no effect.

Three of carbon dioxide's absorption bands are already at maximum absorption, and the fourth competes with water vapor. Taking this fact into account, simple arithmetic shows that doubling the current amount of carbon dioxide in the atmosphere should produce less than 0.6 degrees C of warming. Doubling it again would result in a far smaller increase of temperature. The IPCC, however, estimates much more warming than this from GCMs which are programmed using positive water vapor feedback to greatly amplify the warming from carbon dioxide. Because the net effect of clouds on global temperature is uncertain, water vapor feedback might turn out to be negative, causing global warming theory to crumble.

NOTES AND SOURCES

(1) This quotation is on page 67 of Al Gore's best seller, *An Inconvenient Truth,* Rodale Publishing Company, Emmaus, PA, 2006, 323 pages.
(2) The graph with the incorrect legend from Laura David's book can be viewed here: http://buckhornroad.blogspot.com/2007/09/in-church-of-global-warming-lying-to.html
(3) The second graph from the top at this site shows the 160,000-year Vostok ice core record from drilling in Antarctica: http://www.middlebury.net/op-ed/pangburn.html
(4) The Wikipedia graph mentioned in the text is enlarged here: http://en.wikipedia.org/wiki/File:Atmospheric_CO2_with_glaciers_cycles.gif; the article is here: http://en.wikipedia.org/wiki/File:Atmospheric_CO2_with_glaciers_cycles.gif
(5) *Atmospheric Environment*, Vol. 22, p. 2341, 1988, S.B. Idso, "Carbon dioxide and climate in the Vostok ice core.
(6) *Nature*, Vol. 399, p. 429, 1999, J.R. Petit et al, "Climate and atmospheric history of the past 420,000 years from the Vostok ice core, Antarctica.
(7) *Science*, Vol. 299, p. 1728, 2003, Nicolas Caillon et al, "Timing of atmospheric carbon dioxide and Antarctic temperature changes across Termination III."
(8) *Geophysical Research Letters*, Vol. 32, L19809, 8 October 2005, R.B. Philipona et al, "Anthropogenic greenhouse forcing and strong water vapor feedback increase temperature in Europe."

(9) RealClimate's discussion of water vapor feedback is here:
http://www.realclimate.org/index.php?p=142

(10) 650,000-year ice core record, Antarctica, showing temperature (red) and two greenhouse gases and carbon dioxide (blue): http://en.wikipedia.org/wiki/File:Co2-temperature-plot.svg

(11) Average Daily Lower Tropospheric Temperature 1979-1998:
http://junkscience.com/MSU_Temps/abs_lt.gif

(12) Scroll down the page for the NCDC surface temperature data:
http://junkscience.com/Greenhouse/Greenhouse_not_a_problem.html

(13) *Ibid*, #3

(14) *Ibid*, #3

(15)Chart showing the electromagnetic spectrum:
http://www.lbl.gov/MicroWorlds/ALSTool/EMSpec/EMSpec2.html

(16) Absorption bands of various greenhouse gases in the atmosphere:
http://www.junkscience.com/Greenhouse/absorbspec.gif

(17) An introduction to logarithmic scales is here:
http://en.wikipedia.org/wiki/Logarithmic_scale

(18) Graph estimating clear sky greenhouse effect from quadrupling carbon dioxide:
http://www.junkscience.com/Greenhouse/co2greenhouse-X4.png

(19)http://www.junkscience.com/Greenhouse/co2greenhouse-X4.png

(20) *Ibid*, #18

(21) *Ibid*, #19

CHAPTER TEN—MELTING ICE AND SEA LEVEL RISE

How much will sea level rise if the Arctic ice cap melts completely?

It won't rise at all, not even a single millimeter. The reason is that the Arctic ice cap is frozen sea water, in other words, ice floating on Arctic ocean water. Melting it does not raise sea level just a melting ice cube in a cold drink does not raise the level of the drink. This is also true of an *ice shelf,* such as the Larsen B ice shelf in Antarctica. Although ice shelves consist of glacial ice, which does raise sea level when it melts, the glacial ice in ice shelves has pushed from land out into the ocean. Since the ice is already floating in the ocean, there is no additional rise in sea level when it melts. That already occurred when the ice entered the sea.

The reason melting glacial ice raises sea level is that glaciers only form on land. They originate from fresh water in the form of snow in areas that are above the *snowline,* whether due to high polar latitude or high altitude. In such areas, snow remains on the ground year round. As time goes by and more snow is added each year, layering in the snow is produced from differences in chemistry and texture. Summer snow is distinct from winter snow. A great thickness of snow eventually builds up subjecting the deeper (and older) snow at the bottom to crushing pressure, which causes an increase in density as the air is squeezed out and a gradual conversion into ice. The *stratigraphy* (layering) resulting from this is preserved in the resulting glacial ice.

Whether confined to high mountain valleys, or splaying out from a center and eventually growing large enough to cover an entire continent, such as the ice sheets in Antarctica, all glaciers flow. Mountain glaciers flow down valleys, under the influence of gravity, but ice sheets spread out from their center accumulation area, away the from higher pressure where the ice is thicker. When glacial ice enters water deep enough to float the ice, the upward buoyancy forces the end of the glacier up. The resulting upward stress eventually fractures the end of the glacier, causing it to break away from the main mass. When this happens, the chunk of ice collapses, sometimes producing spectacular geysers of water. Although global warming alarmists love to show the process, known as *calving,* in their books and movies, to make the public think that glaciers are breaking up and rapidly melting because of global warming, the process has nothing to do with climate change. Calving is a normal glacial process and always happens whenever any glacier flows into a body of water large enough to float the ice. It occurs whether the glacier is expanding (advancing) or melting back (retreating). Keep that in mind when viewing pictures showing blocks of ice breaking off and collapsing into the water or when someone in the media goes on about a block of ice the size of Rhode Island breaking off from the Antarctic ice sheet.

Isn't it true that if some of the glacial ice in Greenland and Antarctica melted, or slipped into the sea, sea level would rise by twenty feet?

Whenever glacial ice on land melts, the fresh water, stored as ice, eventually makes it way back to the ocean. This circulation of water is known as the *hydrologic cycle*. It involves evaporation of water from the continents and oceans, precipitation as rain and snow and steam flow. Water, in its various forms, is continually circulating and moving. The main natural process that temporarily removes it from the hydrologic cycle, preventing its circulation, is the formation of glacial ice. When glaciers grow in many areas at the same time, such as in an ice age, enough water is locked up and removed from circulation to cause sea level to fall worldwide. Ice ages, however, eventually end and glaciers melt, returning the stored water to the hydrologic cycle and to the oceans. This restores the water originally lost and sea level worldwide increases.

Isn't it true that sea level is rising, and that this is due to rapid melting of glaciers all over the world because of global warming?

That is partly right—sea level is rising. It started rising about 20,000 years ago as the climate changed into the current interglacial. The climate began to warm and the glaciers started to melt.[1] As the upper graph for Figure 49 clearly shows, most sea level rise occurred during the first few thousand years of the interglacial, but it still continues today at a greatly reduced rate.[2] Before that, expansion of the glaciers caused a tremendous fall in sea level, about 140 meters in fact, which is more than 400 feet.

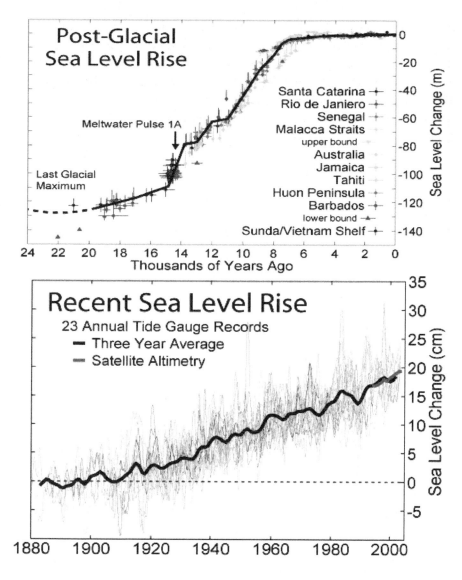

Figure 49. Post glacial and recent sea level rise.

The lower graph shows that total sea level rise during the twentieth century was about 17 cm or 7 inches.[3] That's an average for the entire globe, which can be misleading because what is happening with respect to sea level along a particular coast is strongly affected by local geologic conditions. Even when global sea level is falling during an ice age, sea level can rise locally or fall locally as glaciers retreat during an interglacial. If the land is being uplifted at a fast enough rate, local sea level falls, and if it sinks, sea level rises. The large scale global trend can be overwhelmed by local conditions, which must always be considered when trying to determine what might be happening to sea level on a worldwide scale.

Hasn't the rate of sea level rise accelerated greatly? Al Gore talks about 20 feet in this century.

Global warming alarmists love to sensationalize sea level rise. Al

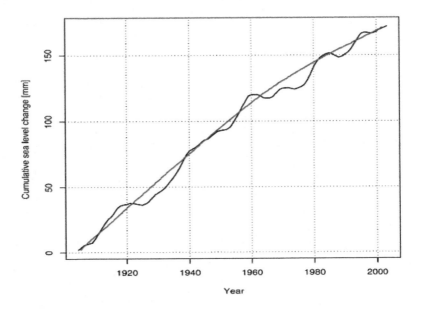

Figure 50. Tidal gauge study showing twentieth century sea level rise

Gore talks about 20 feet and shows pictures in his book with the south part of Florida under water and San Francisco and New York flooded. Based on recent past trends, which is what we should be concerned with rather than climate models, because in science real data must always trump forecasts, there is simply no basis for such a projection. For example, the results of a 2007 peer-reviewed study of tidal gauge records, shown in Figure 50, found no increase in the rate of sea level rise in the late twentieth century.[4] Since 1992, satellite altimetry has provided a much more accurate record of sea level than is available from tidal gauge records. The graph of the satellite data[6] (Figure 51) shows a slower rate of sea level rise during recent years compared to the several years before that.

152

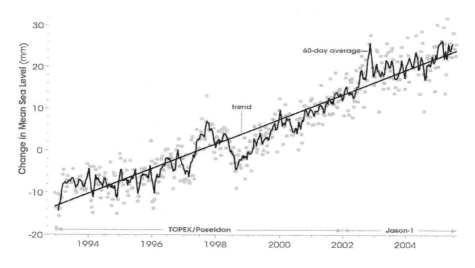

Figure 51. Satellite altimetry record showing recent sea level rise.

Of course, as discussed in Chapter Six, general circulation computer models are not based on past trends, and the IPCC uses them exclusively for their forecasts. Even so, IPCC predictions call for much less sea level rise by 2100 than Al Gore continues to talk about, as shown in Figure 52.[7] Their worst-case scenario shows a sea level rise of only about a meter, about 18 inches, so one wonders where Gore gets his figure of 20 feet.

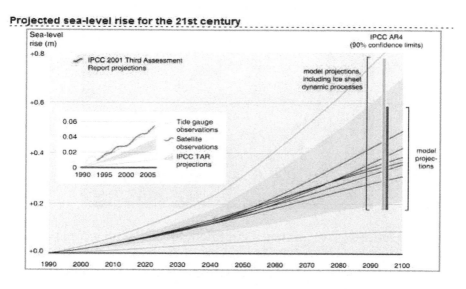

Figure 52. IPCC projected sea level rise for the twenty-first century.

Al Gore's book has satellite images of Greenland showing much more extensive melting in 2005 compared to just a few years earlier. And he cites studies from 2006 claiming that the East Antarctica ice shelf is loosing volume. Isn't it true that glaciers all over the world are melting and that ice is also rapidly melting in Greenland and Antarctica? If so, it also follows that sea level has to increase. Isn't that correct?

If all the remaining ice on the earth were to melt, sea level would rise by about 200 feet.[8] In spite of the media, there is no evidence that this has already started to happen. In fact, most of the evidence is the opposite. For example, the media missed no chance to ballyhoo the seasonal melt back of Arctic ice during the summer of 2008 with tales of woe and pictures of "stranded" polar bears. There were even stories of how soon the fabled northwest passage might become a reality. Somehow, though, they missed the expansion of the ice that started that winter and is still under way. This will be discussed more fully later in the chapter.

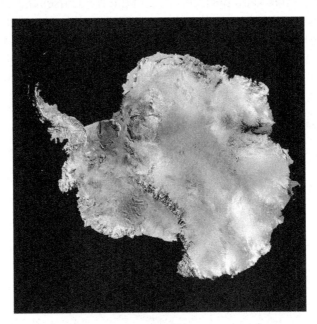

Figure 53. Antarctica, a continent locked in an ice age.

As introductory college geology textbooks make clear, *ninety percent* of the world's ice is in one place, Antarctica. What happens there is really what matters in terms of sea level, so that's the place to start. The first thing to look at is whether Al Gore and other global warming alarmists are right about the ice in Antarctica loosing volume. It seems that whenever the media does a story on Antarctica, it's about the Antarctica peninsula, which looks like an extended finger on the 's satellite view. [9] This area represents only a small proportion of Antarctica's surface area, so one wonders why so many global warming stories center on this region.

This is not a hard question to answer. It's because the peninsula is the only area in Antarctica where the ice is melting. The melt back on the peninsula is clearly documented in a series of satellite images during the Antarctic summer shown in Figure 54 to the left from 31 January 2002 (top) to 5 March 2002 (bottom).[10] The area shown in these images is located in the northern region of the peninsula where the climate is much milder than in the interior of the continent, the area Antarctic veterans refer to as the "banana belt."

Alarmists suggest ice loss in this area is due to a warming climate. Is it? In 2015, a study discovered the real cause — previous unknown underground volcanic activity is warming the crust.

Stories highlighting ice loss in the Antarctica peninsula strongly suggest ice loss over the entire continent. Is this true? Is the rest of Antarctica experiencing the same warming as the peninsula and how important is the loss of ice on the peninsular compared to Antarctica as a whole? These are the key questions.

A research group reported results of their study of temperature trends over mainland Antarctica at the annual meeting of the American Association for the Advancement of Science (AAAS) in 2007. They seemed disappointed at having to say that they found no warming over the past 50 years.[11] Research spokesman David Bromwich of the Byrd Polar Research Center said, "It's very hard in these polar latitudes to demonstrate a global warming signal." Indeed it is, when the mainland is actually cooling and has been for years, in stark contrast to what the computer climate models predict and what Al Gore says. Maybe political correctness kept Dr. Bromwich from drawing such a stark conclusion. "The best we can say right now is that the climate models are somewhat inconsistent with the evidence that we have for the last 50 years from continental Antarctica."

Numerous peer-reviewed studies have found cooling in various areas of the Antarctica mainland since the 1960s. Among them is the remarkable 1.26 degrees F cooling trend in the McMurdo Dry Valleys[12] and the cooling of the central interior.[13] Another study reporting records from 21 surface weather stations and satellite data, found that for Antarctica as a whole, temperatures had declined almost one and a half degrees F.[14]

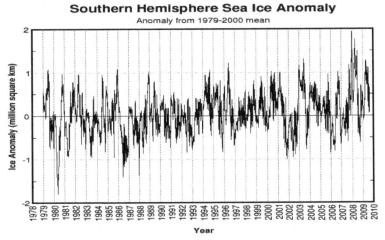

Figure 55. Increase in Antarctic sea ice area, 1978-2009.

Although the media choose not to report it, ice is expanding on the Antarctic continent, not melting. This is true of both ice on the land and on the sea. The British Antarctic Survey recently reported an increase in Antarctic sea ice, but remaining politically correct, went on to say that loss would occur in the future.[15] A graph showing Antarctic sea ice extending back to 1979 (Figure 55) shows record sea ice in 2008 and a generally increasing trend since 1986.[16]

The thickness of the ice is also expanding on the mainland of Antarctica. Satellite altimetry measurements found that the ice sheet covering eastern Antarctica increased in mass by an average of 45 billion tons each year from 1992 to 2003, or nearly half a trillion tons in ten years.[17] Another study a year later reported on an area covering 72% of the Antarctic ice sheet, and found that the ice grew an average 5 mm thicker per year from 1992-2003, or 27 billion tons per year.[18]

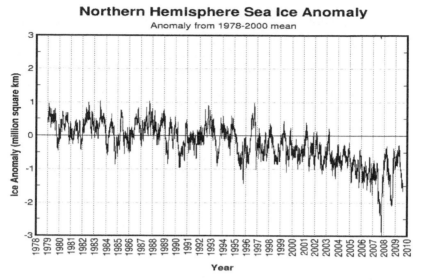

Figure 56. Arctic sea ice area, 1978-2009.

What about the Arctic where Al Gore says "...the most dramatic impact of global warming in the Arctic is the accelerated melting. Temperatures are shooting upward there faster than at any other place on the planet."[19] How does this sensationalized declaration relate to what's been actually happening in the Arctic? The graph of the Arctic sea ice area[20] (Figure 56) shows a slight decrease in ice from 1978 to 1996 and than a steepening trend culminating in the record low sea ice extent of 2007. Then, however sea ice started to increase, and by September 2009, a half million square km had been added. Arctic ROOS provides a daily update of arctic ice on their web site (Figure 57)[21]

Al Gore gives a sensationalized account of rapidly melting Greenland glaciers in his book and movie, and the media ran many feature stories on this. What is actually happening? Using ice core data, a peer-reviewed 1998 study found Greenland temperatures more than 40 degrees F colder than today during the Pleistocene Ice Age. However, the temperature was more than 4.5 degrees F warmer 4,000 to 7,000 years BP during the Climatic Optimum, and nearly 2 degrees F warmer during the Medieval Warm Period.[22] A study using data from the three weather stations with the longest temperature records found "summer temperatures, which are most relevant to Greenland ice sheet melting rates, do not show any

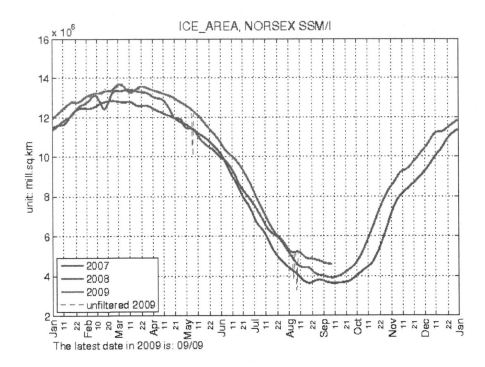

The latest date in 2009 is: 09/09

Figure 57, Arctic ROOS sea ice area in the northern hemisphere.

persistent increase during the last fifty years."[23] Such temperature trends do not fit the rapidly disappearing glaciers in Greenland that Al Gore wrote about.

Was Gore right? Using 10 years of satellite radar altimetry data, a peer-reviewed study of changes in the ice mass in Greenland concluded that, "...the Greenland ice sheet is thinning at the margins and growing inland with a small overall mass gain."[24] This is just one study of many trying to determine what's happening with Greenland's glaciers. Some support this result and some don't. With such conflicting results, it is difficult to reach definitive conclusions about the Greenland ice sheet. Considering that the Greenland ice sheet accounts for only 10% of the world's ice, one wonders why Gore focused so much attention here.

There are many smaller ice sheets in the world and mountain glaciers. Al Gore says on p. 48 of his book, "Almost all of the mountain glaciers in the world are now melting, many of them quite rapidly." The book includes a number of dramatic pictures to prove the point. Is this true?

It's a difficult question to answer because detailed measurements are available for only about 300 of the world's mountain glaciers and only 50 or so have any time series data. The most recent study highlights these inadequacies and concludes that in most areas, mountain glaciers are receding and that this started at the end of the Little Ice Age.[25] A recent study of Himalayan Glaciers found they also began retreating in the mid-eighteenth century late in the Little Ice Age, not in the mid or late twentieth century.[26]

Of course, what is important in terms of global sea level rise is the over all mass balance of all the world's ice. Unfortunately, such data is not available. There is a record of global sea ice since 1979, but since such ice is already floating in the water, melting it cannot contribute to sea level rise. Because global sea level rise during the twentieth century was only a few inches, any contribution from melting glacial ice on land would have to be small. This is especially true considering that water expands as it warms. More than half of the sea level rise during the twentieth century can be attributed to such expansion.[27]

SUMMARY

Glacial ice consists of fresh water and is formed by pressure converting snow into ice in areas above the snowline where snow remains on the ground year round. Glaciers always originate on land, but the edges and ends may push out to sea where the buoyancy of the water forces the ice upward eventually causing the end of the glacier to break and collapse into the water. The process, known as calving, is a normal part of glacier development and occurs whether the ice is expanding or melting back. Spectacular images of calving are used to make the uninformed think it is due to global warming, but this is false.

When glaciers grow on a large scale, such as during the Pleistocene Ice Age, a huge amount of water is removed from the hydrologic cycle, causing sea level to fall worldwide. During the latest glacial advance in the Pleistocene, sea level was 400 feet lower than today. When climate warms during the interglacials, the ice melts and sea level recovers. As part of the current interglacial, this process started about 20,000 years ago and still continues today, but at a much reduced rate compared to the first several thousand years. Sea level has risen about 300 feet during this time, and there is still enough ice in the world to produce another 200-foot rise, if it should all melt, an extremely unlikely event.

The melting of ice shelves, ice floating in sea water, or any situation of floating ice in the ocean, cannot produce a rise in sea level just as melting ice cubes floating in a cold drink cannot raise the level of liquid in the glass.

Specific geologic conditions affect sea level locally. If the land is undergoing geologic uplift, local sea level falls; conversely, if it sinks locally, sea level rises. Local conditions may overwhelm the global trend and must always be taken into account in trying to assess what might be happening to sea level on a global scale.

Global sea level increased during the twentieth century about seven inches. A much safer bet for this century is that it will rise about the same amount, not the 20 feet Al Gore talks about or even the smaller amounts the IPCC predicts based on computer climate models.

In spite of media hype and spin, there is no evidence that the melting of polar ice is accelerating. The area of Arctic ice is expanding again and numerous studies of Antarctica find that overall, the continent is cooling and the thickness of its ice is growing. In spite of Al Gore's exaggerations, New York City is not in danger of drowning.

NOTES AND SOURCES

(1) The U.S. Geological has a good discussion of sea level rise associated with the current interglacial: http://3dparks.wr.usgs.gov/nyc/morraines/flandrian.htm
(2) Graph showing sea level rise in the current interglacial:
http://upload.wikimedia.org/wikipedia/commons/1/1d/Post-Glacial_Sea_Level.png
(3) Graph showing twentieth century sea level rise:
http://upload.wikimedia.org/wikipedia/commons/0/0f/Recent_Sea_Level_Rise.png
(4) *Geophysical Research Letters*, Vol. 34, No. 10, p. 1029, 2007, S.J. Holgate, "On the decadal rates of sea level change during the twentieth century.
(5) Graph showing sea level rise during the twentieth century:
http://www.co2science.org//articles/V10/N4/C1.php
(6) Satellite altimetry record of global sea level:
http://en.wikipedia.org/wiki/File:The_Rising_Sea_Level.jpg
(7) Graph, IPCC projected sea level rise by 2100:
http://maps.grida.no/go/graphic/projected-sea-level-rise-for-the-21st-century
(8) A recommended web site on this subject, because of its thorough documentation, is: http://www.johnstonsarchive.net/environment/waterworld.html
(9) Composite satellite image of Antarctica:
http://en.wikipedia.org/wiki/File:Antarctica_6400px_fromBlue_Marble.jpg
(10) NASA satellite images, Larsen-B ice shelf retreat:
http://svs.gsfc.nasa.gov/vis/a000000/a002400/a002421/index.html
(11) http://www.physorg.com/news90782778.html
(12) *Nature* (advance online publication), January 13, 2002 (DOI 10.1038/nature 710), P.T. Doran et al, "Antarctic climate cooling and terrestrial ecosystem response."
(13) *Science*, Vol. 296, p. 895, 2002, D.W.J. Thompson & S. Solomon, "Interpretation of recent Southern Hemisphere climate change."
(14) *Journal of Climate*, Vol. 13, p. 1674, J.C. Comiso, "Variability and trends in Antarctica surface temperatures from in situ and satellite infrared measurements."

(15) A press release of the British Antarctica Survey's 2009 report on sea ice is here: http://www.antArctica.ac.uk/press/press_releases/press_release.php?id=838
(16) Antarctic sea ice record since 1979: http://Arctic.atmos.uiuc.edu/cryosphere/IMAGES/current.anom.south.jpg
(17) *Science*, Vol. 308, no. 5730, p. 1898, 2005, Curt H. Davis, et al, "Snowfall-driven growth in East Antarctic ice sheet mitigates recent sea level rise."
(18) *Philosophical Transactions of the Royal Society A*, Vol. 364, p. 1627, 2006, D.J. Wingham et al, "Mass balance of the Antarctic ice sheet."
(19) *An Inconvenient Truth*, Rodale Publishers, Emmaus, PA, 2006, p.126, Al Gore.
(20) Graph showing Arctic sea ice anomaly: http://arctic.atmos.uiuc.edu/cryosphere/IMAGES/current.anom.jpg
(21) Daily update of Arctic sea ice area by Arctic ROOS: http://arctic-roos.org/observations/satellite-data/sea-ice/ice-area-and-extent-in-arctic/?searchterm=Sea%20Ice%20Daily%20updated%20time%20series
(22) *Science*, Vol. 282, p. 268, 1998, D. Dahl-Jensen et al, "Past temperature directly from the Greenland Ice Sheet."
(23) *Climatic Change*, Vol. 63, p. 201, 2004, P. Chylek & G. Lesins, "Global warming and the Greenland ice sheet."
(24) *Journal of Glaciology*, Vol. 51, p. 509, 2005, H.J. Zwally et al, "Mass changes of the Greenland and Antarctic ice sheets and shelves and contributions to sea-level rise."
(25) *Progress in Physical Geography*, Vol. 30, p. 285, 2006, R.G. Barry, "The status of research on glaciers and global glacier recession: a review."
(26) *Current Science*, Vol. 96, p. 703, 2009, R.K. Chaujar, "Climate change and its impact on the Himalayan glaciers - a case study of the Chorabari glacier, Garhwal Himalayas, India.
(27) *Nature*, Vol 453, p. 1038, 2008, Catia M. Dominques et al, "Improved estimates of upper-ocean warming and multi-decadal sea-level rise.

CHAPTER ELEVEN—DROUGHTS, FLOODS, EL NINO, WILDFIRES AND HEAT WAVES

The number of weather related disasters taking place is increasing and can be linked to global warming. For example, Al Gore's book shows pictures of Lake Chad drying up. Just forty years ago, it was as big as Lake Erie. Isn't it true that such events will worsen if we don't do something about global warming?

Well, it's certainly true that global warming alarmists constantly talk about such disasters becoming more frequent and more severe. In fact, they say we are already seeing the beginning of this. Al Gore has so testified to Congress on more than one occasion. They blame every weather related disaster on global warming, even unusual cold. For example, during the winter of 2008-2009, London had a snowstorm greater than most people could remember. This was after the Met Office had forecast a warmer than usual winter, but instead, it was the coldest in 30 years. Following the snowstorm, Dr. Myles Allan, a climate modeler at the Met Office, said, "Such snowfalls are in accordance with climate model projections."[1]

Of course, people always talk about the weather, and they always think the most recent extreme event is the worst ever, but is this true? What do studies actually say about such extreme events, such as droughts? Are they actually happening more often and becoming more severe?

A peer-reviewed article published in 2008 used a dataset generated by a computer model to investigate global soil moisture and drought characteristics for the second half of the twentieth century. The study found, "an overall increasing trend in global soil moisture, driven by increasing precipitation, underlies the whole analysis, which is reflected most obviously over the western hemisphere and especially in North America." The authors went on to say, "trends in drought characteristics are predominantly decreasing."[2] Such a study should rank high with global warming alarmists since they place such great faith in computer models, so it seems strange it didn't draw more media attention.

How about some data on actual drought trends rather than computer projections. What do they show?

An important study in 2007 examined severe global droughts from 1901-2000.[3] Among them was the famous Dust Bowl of the Midwest in the 1930s, the Sahel drought in Africa during the 1960s, southeastern Australia in the late 1930s and the northeastern China drought of the 1920s. Twenty-two of the thirty most severe and persistent droughts took place during the first six decades of the century, but only eight occurred during the final four decades, and just three during the last two decades. This is exactly the reverse of the time distribution that would be expected from what global warming alarmists assert. Nevertheless, in his testimony of March 21, 2007 to the U.S. Senate's Environment and Public Works Committee, Al Gore said that "droughts are becoming longer and more intense." He offered no supporting evidence, and no one on the committee demanded any.

Another important peer-reviewed drought article was published in 2007. This one studied droughts in North America over the past thousand years and found evidence for several "megadroughts over the past millennium that clearly exceed any found in the instrumental records."[4] The authors conclude that all recent droughts, including the Dust Bowl, are insignificant when compared to, "an epoch of significantly elevated aridity that persisted for almost 400 years over the AD 900-1300 period."

Warmer air can hold more moisture which means more downpours and more floods. Al Gore said the Midwest floods in June 2008 were due to global warming. The rains there have been heavy again this year. On page 106 of *An Inconvenient Truth*, a graph shows that "the number of large flood events has increased decade by decade, on every continent." Isn't this our future if we don't act?

Sure, global warming alarmists say there will be more floods because all the GCMs forecast an enhanced hydrologic cycle and increased flooding. That doesn't stop them from arguing that global warming will increase both droughts and floods. After all, they blame every extreme weather event on carbon dioxide and global warming, even the fact that Australia had its coldest winter in 60 years during 2008.[5]

As for the graph in Gore's book showing 70 major floods in the 1950s and more than 600 in the 1990s, it is impossible to know what this graph is actually showing because Gore never defines what constitutes a "major" flood. He does list the *Millennium Ecosystem Assessment* as the source of the graph. Published by a group sympathetic to global warming alarmism with major funding from the United Nations, its data and claims are as suspect to skeptics as those in Exxon publications are to alarmists. Going to the report itself and the flood graph on page 119 does not help in trying to understand how the group defines a "major" flood.[6] It appears, based on the huge increase the graph shows, that a major flood is defined by the monetary damage the flood causes, with no adjustment for inflation. If so, the graph is misleading because what $1 would buy in 1950 now takes $8.91.[7] In other words, because of inflation, the value of the U.S. dollar has declined by a huge amount since 1950.

Reliable data is available on flooding and major floods in peer reviewed journal articles. One such article studied river flow data for a group of 21 stations with long term records in many different parts of the world. Individual record lengths varied from 44 to 100 years and averaged 68 years. The authors' analyses showed that more than half of the stations had reduced flooding, not increased.[8] Including that sort of data in *An Inconvenient Truth* would have greatly improved it, but of course doing so did not fit Gore's preconceived position.

From a geological viewpoint, V.R. Baker's work on floods is of particular significance. In a 2004 peer-reviewed publication,[9] he points out that two very different methods exist for assessing the risk of future flooding. One method uses computer models and claims that extreme floods may increase because of global warming. In contrast, the method he advocates determines how flooding in the past responded to changes in climate and assesses the risk of future flooding based on that knowledge. His analyses of past floods show that "the floods of recent years do not generally exceed ... those of past clusters, and much larger floods are usually indicated in the past...."

Even if more flooding were occurring, that does not mean warming caused it. Many other factors play a role in causing floods besides climate, especially reducing forest cover, increasing urbanization, paving large areas and channelizing streams.

What about El Niños? Aren't they predicted to become more destructive and to develop with greater frequency?

Indeed, according to GCM predictions, they will develop more often and will be more powerful with greater destructive potential. If correct, more weather-related disasters related to ENSO events (El Niño-Southern Oscillations, see Chapter Six) await us in the future. With computer generated forecasts, the question always is how accurate is the projection likely to be. There are many palaeoclimate studies that suggest they are not very accurate. In fact, such studies tend to find the exact opposite, that past El Niño events were less frequent during times of warmer climate and happened more often during cooler times, such as the Little Ice Age.

For instance, a peer reviewed study from 2002 that analyzed ENSO events from 1607-1990, found the period from 1820 to 1860 to have been particularly active.[10] An earlier study that used both instrument records and proxy data found persistent and long-lasting ENSO events in both the eighteenth and nineteenth centuries, when the temperature was lower.[11] To cite just one more example of ENSO research, a 2003 peer-reviewed study used oxygen isotope proxy data to study the frequency and intensity of these events back to the mid Holocene.[12] The results found that ENSO events were "considerably weaker or absent between 8.8 and 5.8 ka," which was during the Climatic Optimum, the warmest part of the current interglacial.

Based on these and numerous others studies, it would appear that the climate models are wrong again when it comes to El Niño events. They do not seem to have been more intense or to have occurred with greater frequencies in the past during times when the climate was warmer.

Early in 2009, 300 people were killed in Australian brush fires. Not long before that, there was a major wildfire in California, and this fall, another one. Isn't it true that more and more wildfires are occurring as hotter temperatures dry out flammable materials on the ground?

If there were a relation with global temperature, there should be fewer now because, as mentioned previously, the earth has not warmed in several years. Global warming alarmists ignore this and just keep talking about all manner of calamities, including wildfires. Al Gore says wildfires are becoming more common, and shows a graph on p. 229 of *An Inconvenient Truth* to illustrate his point. It shows a tremendous increase in North and South American major wildfires from 1950 to 2000. As true of floods, "major" is not defined, and since the source of this data is the same previously mentioned regarding major floods, the *Millennium Ecosystem Assessment*, going to the original sheds no light on the matter. If "major" refers to a certain monetary value of property loss, say a million dollars, because today's dollar buys 8.91 less than the 1950 dollar,[13] many more fires today would reach that level than in 1950. Even if an inflation adjustment had been used, the data would still be skewed toward more wildfires because there are many more homes and buildings now than in 1950.

The media considers it common knowledge that wildfires are happening more often and, since they make such spectacular video, they always highlight them. Because, to twist an old saying, the media and Al Gore can't be wrong, this must be true. Some people, however, those who like to think for themselves, still want to see some data. Fortunately, reliable data is available.

A 2008 peer-reviewed study used, "sedimentary charcoal records spanning six continents to document trends in both natural and anthropogenic biomass burning [over] the past two millennia."[14] Results showed that burning increased sharply between 1750 and 1870, and then began to decline. The authors found an association between cooler climate and more fire and attribute the decrease since 1870 to "the global expansion of intensive grazing, agriculture and fire management."

Other studies have found no increase in fires in recent years. One that is particularly interesting analyzed the years between 1981 and 2000 using data from NOAA's Pathfinder satellite. The high resolution instruments on this satellite allow, for the first time, definitive data collection on fires for the entire globe. The data shows that in portions of Eurasia and western North America there was a sharp upward trend in the land area burned, but in southeastern Asia and Central America, downward trends occurred. Overall, "there was no significant global annual upward or down trend in burned area."[15]

The fact that this study was published a year after Gore's book does not excuse his use of the misleading graph because several previous studies were available showing that his concerns were not backed by data. To mention just one, a 2001 peer-reviewed study concluded that a warmer climate "is likely to be less favorable for fire ignition and spread in the east Canadian boreal forest than over the last 2 millennia."[16]

A massive heat wave struck Europe in 2003, killing 35,000 people. In 2005, over 200 records for high temperatures were set in the American west. Is there any better proof that the earth is growing dangerously hot?

Certainly, high temperature records should become more and more common and low temperature records should decrease if global warming is a real problem. Evidence showing that this is the case would have to be counted as favoring a real warming of the earth, while the reverse should be a strike against this. Global warming alarmists consider a single heat wave as evidence of serious warming, but they remain silent about cold snaps, such as the fact that for the first time in sixty years, snow fell in June 2009 in North Dakota.[17] The author of the referenced article draws attention to several areas in the world where crops are under stress because of cool weather and observes, "Our politicians haven't noticed that the problem may be that the world is not warming but cooling."

Extreme weather events are a favorite for television coverage. In the hyperbole that seems to be part of their profession, just about every heat wave is referred to as "record breaking." Of course, neither a single heat wave or cold spell is evidence of anything except that the weather is fickle and changeable. What really counts is documented evidence showing a trend. Such peer-reviewed evidence is available.

An especially interesting 2006 study examined the statistics of record-breaking temperatures in relation to global warming by comparing predictions resulting from a Monte Carlo simulation to the actual weather record for the city of Philadelphia, PA, going back 126 years.[18] As a result of their study, the authors concluded, "the current warming rate is insufficient to measurably influence the frequency of record temperature events." They also say they "cannot yet distinguish between the effects of random fluctuations and long-term systematic trends on the frequency of record-breaking temperatures."

A 2007 peer-reviewed study examined the severe European heat wave of 2003, and found that "soil moisture-temperature interactions increase the heat wave duration and account for typically 50-80% of the number of hot summer days."[19] This means that soil moisture depletion increases the intensity and duration of heat waves. In terms of increasing heat wave intensity, this is especially interesting because seven peer reviewed articles have found that soil moisture had increased over the past half century. [20, 21, 22] This is attributed to the discovery that many plant species use less water when growing in an atmosphere of increased carbon dioxide, leaving more in the soil and therefore contributing to an increase in soil moisture. If this is an important relationship, it seems that increasing carbon dioxide in the atmosphere might make a contribution to reducing the intensity and frequency of heat waves.

Before ending this chapter, a comment on the massive brush fire that killed 300 people early in 2009 in southeastern Australia. Global warming alarmists blame the drought on the heat wave gripping Victoria at the time, but both are not unusual in Australia. Instead of the heat, many loud and vocal Australians blame the large number of deaths on governmental adoption of what they called "green" policies that prevented controlled burns and the clearing of underbrush. Scott Gentle, a resident of a town near the center of the fire where more than 150 people died horribly, warned in testimony to a government body more than a year before the fire: "Living in an area like Healesville, whether because of dumb luck or whatever, we have not experienced a fire since about 1963. God help us if we ever do, because it will make Ash Wednesday look like a picnic."[23] He was right.

SUMMARY

One of the most often cited supposed effects of global warming is that many sorts of destructive weather events will increase in severity and become more common. Al Gore has testified to this before Congress on several occasions. Even so, the number of deaths annually per thousand people from extreme weather events continues to decline.

Global warming alarmists argue that a warmer earth will cause both drought and flood. Their logic is that warmer air can hold more moisture, producing more rain; yet, higher summer temperatures reduce soil moisture, causing drought. For example, Al Gore blamed the Midwest floods of June 2008 on global warming. The actual data, however, shows that U.S. flood damage has been decreasing. The same is true for drought. Princeton University research found in a 2008 study that "drought characteristics are predominantly decreasing," along with "an overall increasing trend in global soil moisture."

Climate models have consistently predicted that global warming will cause destructive El Niños to develop with greater frequency. Palaeoclimate studies find exactly the opposite. Past El Niño events seem to have been less frequent during warmer times in the past and more frequent during colder periods including the little ice age.

The media considers it common knowledge that wildfires are becoming more frequent and more destructive, as exemplified by the Australian fires in early February 2009. Fortunately, data from the NOAA-NASA Pathfinder satellite shows the common knowledge is wrong. The satellite's Advanced Very High-Resolution Radiometer allows us for the first time to monitor wildfires over the entire surface of the earth with heretofore unavailable precision and accuracy. Data from the satellite show no global trend in wildfires from 1981-2000.

An obvious effect of a warming earth would be more severe and more frequent summer heat waves. If the earth warmed significantly, this would have to be true, but peer reviewed studies have shown that the amount of warming that has actually occurred since 1880 is so small its influence on record heat waves cannot be measured. For the United States, the 1930s have more heat records than any other decade.

NOTES AND SOURCES

(1) *The Daily Telegraph*, UK, 3 February 2009, Richard Alleyne, Science Correspondent, "Snow is consistent with warming."
(2) *Journal of Climate*, Vol. 21, p. 432, 2008, J. Sheffield & E.F. Wood, "Global trends and variability in soil moisture and drought characteristics."

(3) *Geophysical Research Letters*, Vol. 34, p. 10.1029, 2007, G.T. Narisma et al, "Abrupt changes in "rainfall during the twentieth century."

(4) *Earth-Science Reviews*, Vol. 81, p. 93, 2007, E.R. Cook et al, "North American Drought: Reconstructions, causes, and consequences."

(5) *Sydney Morning Herald*, September 1, 2008, Cathy Alexander, "Big chill a symptom of climate chaos." The story is available online here: http://news.smh.com.au/national/big-chill-a-symptom-of-climate-chaos-20080901-46yx.html

(6) The Synthesis Report is available as a pdf file here: (Caution - a long download is required) http://www.millenniumassessment.org/en/synthesis.aspx

(7) Inflation calculator for the U.S. dollar is on the web here: http://www.dollartimes.com/calculators/inflation.htm

(8) *Hydrological Sciences Journal*, Vol. 50, p. 811, 2005, C. Svensson & Z.W. Kundzewicz, "Trend detection in river flow series: 2. Flood and low-flow index series."

(9) *Journal of the Geological Society of India*, Vol. 64, p. 395, 2004, V.R. Baker, "Palaeofloods and global change."

(10) *Paleoceanography*, Vol. 17, p. U71, 2002, M.N. Evans et al, "Pacific sea surface temperature field reconstruction from coral O^{18} data using reduced space objective analysis."

(11) *Holocene*, Vol. 8, p. 101, 1999, R.J. Allan & R.D. D'Arrigo, "Persistent ENSO sequences: How unusual was the 1990-1995 El Niño?"

(12) *Geophysical Research Letters*, Vol. 30, 10.1029/2002GL015868, C.D. Woodroffe et al, "Mid-late El Niño variability in the equatorial Pacific from coral microatolls."

(13) *Ibid*, #7

(14) *Nature Geoscience*, Vol. 1, p. 697, 2008, J.R. Marlon et al, "Climate and human influences on global biomass burning over the past two millennia."

(15) *Global Change Biology*, Vol. 13, p. 40, 2007, D. Riano et al, "Global spatial patterns and temporal trends of burned area between 1981 and 2000 using NOAA-NASA Pathfinder."

(16) *Journal of Ecology*, Vol. 89, p. 930, 2001, C. Carcaillet et al, "Change of fire frequency in eastern Canadian boreal forests during the Holocene: does vegetation composition or climate trigger the fire regime?"

(17) *The Sunday Telegraph*, UK, June 14, 2009, Christopher Booker, "Crops under stress as temperatures fall."The story is available online: http://ww.telegraph.co.uk/comment/columnists/christopherbooker/5525933/Crops-under-stress-as-temperatures-fall.html

(18) *Physical Review*, Vol. E74, 061114, 2006, S. Redner and M.R. Peterson, "Role of global warming on the statistics of record-breaking temperatures."

(19) *Geophysical Research Letters*, Vol. 34: 10.1029/32006GL029-068, 2007, E.M. Fischer et al, "Contribution of land-atmosphere coupling to recent European summer heat waves."

(20) *Bulletin of the American Meteorological Society*, Vol. 81, p. 1281, 2000, A. Robock et al, "The global soil moisture data bank."

(21) *Geophysical Research Letters*, Vol. 32: 10,1029/2004GL021914, 2005, A. Robock et al, "Forty-five years of observed soil moisture in the Ukraine: No summer desiccation (yet)."

(22) *Journal of Geophysical Research*, Vol. 112: 10.1029/2006JD007455, 2007, H. Li & M. Wild, "Evaluation of Intergovernmental Panel on Climate Change Fourth Assessment soil moisture simulations for the second half of the twentieth century."

(23) *The Sydney Morning Herald*, February 12, 2009, Miranda Devine, "Green ideas must take blame for deaths." The story is available online at: http://www.smh.com.au/environment/green-ideas-must-take-blame-for-deaths-20090211-84mk.html

CHAPTER TWELVE—SEVERE STORMS AND TROPICAL DISEASES

Something happened in 2005 that had never happened before. So many hurricanes and tropical storms developed in the Atlantic that the World Meteorological Organization ran out of names. They had to start using Greek letters such as Alpha, Beta and Delta for storm names. The explanation seems obvious, global warming.

There's no doubt that 2005 was a big year for Atlantic hurricanes, and the timing was perfect for Al Gore. It's one of the main things he talks about in his book where he outlines the basic theory: "As water temperatures go up, wind velocity goes up, and so does storm moisture condensation."[1] Based on this theory, NOAA's Geophysical Fluid Dynamics Laboratory says in a 2008 report on the relationship between hurricanes and global warming that, "It is likely that greenhouse warming will cause hurricanes in the coming century to be more intense on average and have higher rainfall rates than present-day hurricanes."[2] The NOAA report also says, however, that "It is premature to conclude that human activity--and particularly greenhouse warming--has already had a discernible impact on Atlantic hurricane activity."[3] This is certainly a more cautious stance than Al Gore who argues that we are already seeing more hurricanes of increased destructive power because of human influences on climate.

Why do Al Gore and the NOAA report differ on this key point?

The NOAA report actually looks at the record of past tropical storms and hurricanes in the Atlantic. The author realizes that past reporting of these storms was far more spotty and less reliable than in

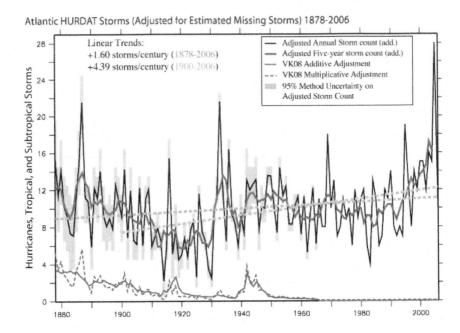

Figure 58. Atlantic storms, 1878-2006, NOAA.

recent years, and attempts to make adjustments by estimating the number of storms not reported. Although Figure 58, a graph of the adjusted data,[4] shows a slight upward trend from 1878 to 2006, "...statistical tests reveal that this trend is so small, relative to the variability in the series, that it is not significantly distinguishable from zero."[5]

From a geological viewpoint, the past is always key to understanding the future, and abundant research is available concerning past hurricanes and tropical storm activity in the Atlantic. Before examining the conclusions of some of this research, a NOAA graph showing named tropical cyclones in the Atlantic from 1851 through 2004 (Figure 59) is useful to examine in order to have a basic understanding of the situation.[6] The yellow bars in the graph show all named tropical cyclones, green all named hurricanes, and the red, severe hurricanes, Saffir-Simpson Category 3 and greater. Only actual named storms are included, with no additions due to an estimated number of missed storms in the old records.

A couple of things are apparent in looking at this data. First, the number of storms is highly variable from year to year, with no obvious recurring pattern. If there is an overall increase in the number of storms, it is not obvious, particularly that quite a few storms were

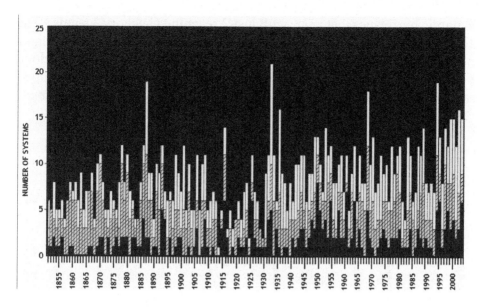

Figure 59. Named tropical Atlantic cyclones, 1851-2004, NOAA.

most likely not included in the earlier data. Considering only the more recent portion of the graph from the end of World War II, when major storms are not likely to have been missed, a trend in severe hurricanes is obvious: few severe hurricanes from the early 1970s to 1995 sandwiched between periods of many such storms before and after. This is suggestive of a cyclic pattern. It is also interesting that 1950 had the greatest number of Category 3, 4 and 5 storms of any year during this period, or the entire record.

Simply taking a few moments to look at the most basic type of data concerning Atlantic hurricanes shows there is no apparent basis of global warming alarmists' contention that hurricanes are increasing in number and severity. This leaves alarmists to fall back on the predictions of general circulation models which predict increasing numbers of hurricanes. It is up to each individual to decide which is more important-- actual data or predictions of computer models.

What if data more recent than 2004 was included? After all, as Gore's book makes clear, 2005 was a record year for hurricanes.

Good point. At another web site, NOAA has a new graph that begins in 1944 and runs through 2008 (Figure 60).[7] The same pattern still holds--a pattern of many hurricanes separated by fewer hurricanes. Notice, the record year for major hurricanes of the kind that cause most of the damage is still 1950. After the record year of 2005, 2006 must have been a real let-down for global warming alarmists. The were only 5

Atlantic hurricanes that year, with just 2 reaching Category 3 status.[8] The following year, 2007, was similar with 6 hurricanes and only 2 major storms, but both at the strongest Category 5 level.[9] Tropical cyclone activity picked up in the Atlantic during the 2008 season with 16 named storms, 8 hurricanes, and 5 of those major, but no Category 5 storms.[10] As of mid-September,

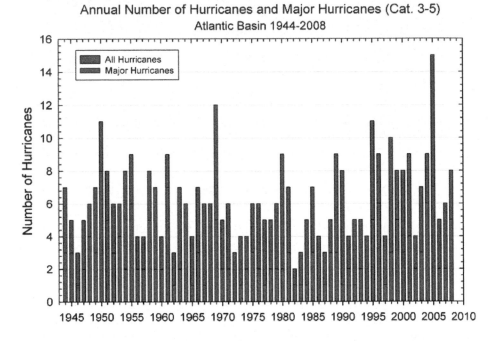

Annual Number of Hurricanes and Major Hurricanes (Cat. 3-5)
Atlantic Basin 1944-2008

Figure 60. Atlantic hurricanes, 1944-2008, NOAA.

only two Atlantic hurricanes have occurred during the 2009 season, suggesting another year of low activity.

Including this more recent data in the graph only serves to reinforce the previously deduced pattern. With respect to Atlantic hurricanes since 1944, there seems to be 25-30-year cyclic pattern, both in the number of hurricanes and the number of major hurricanes. There is little or no relationship between global temperatures and the number of hurricanes.

There's a lot of research saying warming is related to hurricanes. Maybe the most famous example is the MIT study that was published less than a month before Katrina hit. I haven't heard of any peer-reviewed studies proposing hurricane cycles such as you claim to see in the NOAA data.

Al Gore mentions the MIT study on page 92 of *An Inconvenient Truth*, saying, "...a major study from MIT supported the scientific consensus that global warming is making hurricanes more powerful and more destructive." Gore seems to be referring to a letter published in the August 4, 2005 issue of *Nature*. In it, Dr. Kerry Emanuel of MIT says his work indicates that "future warming may lead to an upward trend in tropical cyclone destructive potential."[11] Because of the fortuitous timing of publication, the media made him famous. They ignored the response to Dr. Emanuel's article that pointed out there has been no upward trend in the destructiveness of hurricanes if the effects of societal changes are removed from the data.[12]

More recently, however, Dr. Emanuel had an article in the March 2008 issue of the *Bulletin of the American Meteorological Society*. In it, he states that the variability of Atlantic hurricanes, "is controlled mostly by time-varying radiative forcing owing to solar variability, major volcanic eruptions, and anthropogenic sulfate aerosols and greenhouse gases, though the response to this forcing may be modulated by natural modes of variability."[13] The media also ignored this later article, as might be expected, when natural causes are placed ahead of greenhouse gases as potential controlling factors of hurricane variability.

As for the frequency of hurricanes, there's actually been quite a lot of peer-reviewed research in this area. For instance, a 1998 study examined all landfalling Gulf Coast hurricanes during the one hundred years prior to 1996, and found that instead of more and stronger hurricanes, the trend has been the opposite, fewer and weaker in the more recent decades.[14] A study in 2000 of all landfalling Atlantic basin hurricanes from 1935 to 1999 found a negative slope in their frequency rather than an increase.[15] A few years later, a statistical study of major North Atlantic hurricanes during the twentieth century found an increase in frequency since 1994 with an average of 3.86 during the 7-year period from 1995 through 2001, but less than the 4.14 during the 14-year period 1948-1961, so there was no overall increase.[16] A more recent study calculating the return periods of landfalling Atlantic hurricanes included the very active 2005 hurricane season. In years, the return periods ranged from 0.9 year for Category 1 to 23.1 years for Category 5, and the strike period from 1900 through 2005 had not increased,[17] despite what Al Gore and other alarmists say.

Some research has found that Atlantic hurricane activity is cyclic. A study published in the *Journal of Geophysical Research* found a 60-year quasi-periodic cycle in hurricane activity and that "if there is an increase in hurricane activity connected to a greenhouse gas induced global warming, it is currently obscured by the 60-year quasi-periodic cycle."[18] Another 2008 peer-reviewed article also found a multidecadal periodicity in Atlantic hurricane activity with two minimal periods, 1904-1929 and 1970-1994, and three maximal periods, 1880-1904, 1929-1970 and 1994-present.[19]

Assuming such a recurring pattern actually exists, has anyone speculated on what might be causing it?

It has been related to various things such as El Niño Southern Oscillation, the 70-year Atlantic multidecadal oscillation,[20] and the much shorter North Atlantic Oscillation.[21] Two Florida State scientists found an unexpected relationship between times of decreased solar activity (fewer sunspots) and increased Atlantic hurricane frequency.[22]

Hurricanes or tropical cyclones happen in many parts of the world, not just the north Atlantic. Is there a relation between warming and such storms on a global scale?

Research on this question has produced mixed results. As already mentioned, Emanuel in 2005 found that the tropical cyclone power dissipation index in the Atlantic and northwest Pacific has increased.[23] Webster studied data from all tropical cyclone basins and found almost twice the number of Category 4 and 5 storms over a 1990-2004 period as compared to 1975-1989.[24] Klotzbach questioned that conclusion the following year, challenging the quality of the data. Using a "near-homogeneous" global data set for the years 1986-2005, he concluded that "no significant change in global net tropical cyclone activity" has occurred over the studied time period.[25] These results are supported in a 2007 study using satellite data; the five authors nicely summed up the situation in their conclusion saying, "the question whether hurricane intensity is globally trending upwards in a warming climate will likely remain a point of debate in the foreseeable future."[26]

For more information on hurricanes, an especially useful website is provided by ICAT, an insurance underwriter. The interactive site shows the tracks of all hurricanes in the north Atlantic since 1900, the amount of damage adjusted for inflation and the category.[27] An introductory video shows how to use this fascinating site.

New diseases are emerging and tropical diseases once under control are spreading into new areas. A warming climate is blamed. How dangerous is the threat?

This is one of the major scare tactics that global warming alarmists use. Al Gore shows microscopic photos of a dozen different microbes with a caption that reads, "Some 30 so-called new diseases have emerged over the last 25 to 30 years. And some old diseases that had been under control are now surging again."[28]

Perhaps malaria is the best known disease used as an example of how global warming is supposed to cause tropical diseases to spread. Not many people realize that in the eighteenth and nineteenth centuries, when the climate was cooler, malaria was common in Europe and Russia, and it was found all over the United States, as Figure 61 makes clear.[29] Today, however, if a case occurred it would make the national news and some expert would be blaming it on global warming. Malaria's decline was due to less direct contact between insects and humans, improved public health, improved living conditions, better insect control and better treatments, not climate change.

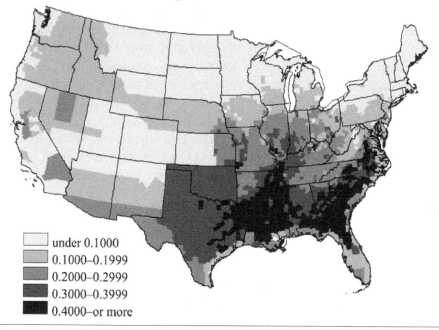

Figure 61. Malaria occurrence in the late nineteenth century.

A particularly comprehensive peer-reviewed study of malaria and climate was published in 2000. The scientists used the present-day distribution of the disease to determine its climate constraints and applied this to GCM predicted future climates. Their results, "contradict[s] prevailing forecasts of global malaria expansion."[30]

Several types of mosquito-borne diseases are also supposed to increase because of global warming. As Al Gore says, "...mosquitoes are profoundly affected by global warming."[31] Well, he is right, but not in the way he means, if the results of an interesting 2003 study are found to be generally applicable. The study involved feeding leaf litter from the same species of poplar tree to four different species of mosquito larvae and noting their rate of development. The leaves from one group of trees were grown in air with 360 ppmv carbon dioxide and the other, 720. The interesting results were that are all four types of mosquitoes developed much more slowly on the leaf litter from the high carbon dioxide atmosphere.[32] The important implication is that a longer development time for the larvae is likely to mean fewer mosquitoes to carry disease, therefore less disease.

SUMMARY

One of the most often repeated litanies from global warming alarmists is that a warming planet means stronger and more frequent hurricanes that cause more and more destruction. According to NOAA's Geophysical Fluid Dynamics Laboratory, "even more intense hurricanes over the next century" may be expected "as the earth's climate is warmed by increasing levels of greenhouse gases in the atmosphere." Predictions such as this regarding hurricanes and other severe storms result from general circulation computer models. Al Gore implies in his book that we don't have to wait for the next century, that we can already seeing this happening.

Other than evaluating the reliability of GCMs, as discussed in Chapter Six, there is no way to judge how accurate such forecasts might be, but we can look for trends in past records of hurricanes. This is not as straight forward as it might seem because reports of past hurricanes were far more spotty than today, particularly for those not striking a coast. A routine procedure is to adjust older data to allow for this, but this is little better than guess work because there is no way to actually know how many storms were not counted. The record is, however, thought to be reliable from the 1940s. Simply inspecting a graph showing the number of Atlantic hurricanes since then is informative. The data does not show more hurricanes or that hurricanes are becoming more intense.

A number of peer-reviewed studies have reached this same conclusion. Among them is a 2008 study published in the *Journal of Climate* that agreed with numerous previous studies in finding no trend in Atlantic hurricanes since the nineteenth century. For the earth as a whole, a 2007 study published in *Geophysical Research Letters* using satellite data from 1983-2005 did not find "upward trends in hurricane intensity." Many other studies have reached the same conclusion.

Some scientists find a cyclic pattern in the occurrence of past hurricanes that might be related to the 70-year Atlantic multidecadal oscillation or the shorter period North Atlantic Oscillation. In a 2008 peer-reviewed article, two Florida State researchers found an unexpected link between times of decreased solar activity (fewer sunspots) and increased hurricane intensity.

Another common lament from alarmists is that tropical diseases will become common in mid-latitude regions such as Europe and the United States because of predicted warming. This is an area where the past can be especially helpful. Few people realize that diseases such as malaria were widespread in the U.S., Europe and Russia during the Little Ice Age when climate was cooler than today. Tropical diseases are rare today in these areas despite warming because of better public health, better insect control and advances in medicine. A surprising result from a 2001 study suggests that increasing carbon dioxide in the atmosphere may reduce the number of mosquitoes, and by implication, the spread of mosquito-borne diseases. Studies can also be cited that have found no correlation between warming and a number of other vector-borne diseases, tick-borne diseases, Dengue Fever and Atopic eczema.

NOTES AND SOURCES

(1) *An Inconvenient Truth*, Rodale Publishing Co., Emmaus, PA, p. 89, 2006, Al Gore
(2) *Geophysical Fluid Dynamics Laboratory/NOAA*, October 17, 2008, Thomas R. Knutson, "Global warming and hurricanes: An overview of current research results." Available on the web at: http://www.gfdl.noaa.gov/global-warming-and-hurricanes
(3) *Ibid*
(4) A graph showing Atlantic tropical storms and hurricanes is here: http://www.gfdl.noaa.gov/pix/user_images/tk/global_warm_hurr/Adjust_TS_Count.png
(5) *Ibid*, #2.
(6) NOAA graph showing Atlantic tropical cyclone activity from 1851 through 2004: http://www.nhc.noaa.gov/gifs/atlhist_lowres.gif
(7) NOAA graph showing Atlantic hurricanes from 1944 though 2008: http://www.ncdc.noaa.gov/img/climate/research/hurricanes/fig1-atlantic-all-and-major.gif
(8)http://en.wikipedia.org/wiki/2006_Atlantic_hurricane_season
(9)http://en.wikipedia.org/wiki/2007_Atlantic_hurricane_season
(10)http://en.wikipedia.org/wiki/2008_Atlantic_hurricane_season
(11) Dr. Kerry Emanuel's 2005 letter in *Nature* is available here: inhttp://www.nature.com/nature/journal/v436/n7051/abs/nature03906.html
(12) Dr. Roger Pielke's response to Dr. Emanuels's article is at: http://www.nature.com/nature/journal/v438/n7071/full/nature04426.html
(13 Dr. Emanuel's later article is here:
ftp://texmex.mit.edu/pub/emanuel/PAPERS/Haurwitz_2008.pdf

(14) *Monthly Weather Review*, Vol. 126, p. 1174, 1998, C.W. Landsea et al, "The extremely active 1995 Atlantic hurricane season: environmental conditions and verification of seasonal forecasts."

(15) *Australian & New Zealand Journal of Statistics*, Vol. 42, p. 271, 2000, F. Parisi & R. Lund, "Seasonality and return periods of landfalling Atlantic basin hurricanes."

(16) *Journal of Climate*, Vol. 17, p. 2652, 2004, J.B. Elsner et al, "Detecting shifts in hurricane rates using a Markov Chain Monte Carol approach."

(17) *Journal of Climate*, Vol. 21, p. 403, 2008, F. Parisi & R. Lund, "Return periods of continental U.S. hurricanes."

(18) *Journal of Geophysical Research*, Vol. 113, 10.1029/2008JD010036, 2008, P. Chylek & G. Lesins, "Multidecadal variability of Atlantic hurricane activity."

(19) *Journal of Climate*, Vol. 21, p. 3929, 2008, P.J. Klotzbach & W.M. Gray, "Multidecadal variability in North Atlantic tropical cyclone activity."

(20) A summery of the Atlantic multidecadal oscillation is available here: http://en.wikipedia.org/wiki/Atlantic_Multidecadal_Oscillation

(21) A short explanation of the North Atlantic Oscillation is here: http://en.wikipedia.org/wiki/Atlantic_Multidecadal_Oscillation

(22) The abstract of the Florida State article is available at: http://en.wikipedia.org/wiki/North_Atlantic_oscillation

(23) *Ibid*, #11.

(24) *Science*, Vol. 309, p.1844, 2005, P.J. Webster et al, "Changes in tropical cyclone number, duration and intensity in a warming environment."

(25) *Geophysical Research Letters*, Vol.33, 10.1029/2006GL025881, 2006, P.J. Klotzbach, "Trends in global tropical cyclone activity over the past twenty years (1986-2005)."

(26) *Geophysical Research Letters*, Vol. 34, 10.1029/2006GL028836, 2007, J.P. Kossin et al, "A globally consistent reanalysis of hurricane variability and trends."

(27) The ICAT interactive hurricane web site: http://www.icatdamageestimator.com/

(28) *Ibid*, #1.

(29) The history of malaria is summarized in the U.S. along with a map showing its distribution in the nineteenth century is at this site: http://www.pubmedcentral.nih.gov/articlerender.fcgi?artid=2600412

(30) *Science*, Vol. 289, p. 1763, 2000, D.J. Rogers & S.E. Randolph, "The global spread of malaria in a future, warmer world."

(31) *Ibid*, #1, p. 173.

(32) *Freshwater Biology*, Vol. 48, p. 1432, 2003, N.C. Tuchman, et al, "Nutritional quality of leaf detritus altered by elevated atmospheric carbon dioxide: effects on development of mosquito larvae."

CHAPTER THIRTEEN—ACIDIC OCEANS, CORAL BLEACHING AND POLAR BEARS

A 2008 article in *Discover* magazine said all the extra carbon dioxide in the atmosphere is killing life in the sea by turning the oceans acidic. How does that occur?

Acidification of the oceans has started receiving a lot of publicity in the last few years. For example the magazine article said that, "Industrial carbon dioxide is turning the oceans acidic, threatening the foundation of sea life."[1] The scare is that as carbon dioxide increases in the atmosphere, more is absorbed into ocean water increasing the amount of dissolved carbonic acid (see Chapter Two) and lowering the pH. This is supposed to make it harder for shelled invertebrates to precipitate enough calcium carbonate to build their shells. In addition, if the pH became low enough, the shells of invertebrates could start to dissolve. The magazine article treats this as yet another crisis because the change is, "happening much faster than animals can adapt."

Al Gore also discusses ocean acidification in his book,[2] where he says, "Carbonic acid resulting from all the extra carbon dioxide changes the pH in ocean water and alters the ratio of carbonate and bicarbonate ions." Except for the implication that humans cause the carbonic acid in ocean water, this statement is essentially correct in a controlled laboratory situation, but is overly simple for the oceans due to the buffering action of the calcium carbonate bearing rock limestone and limestone sediments on the sea floor. Scientists think this buffering action maintains the pH of ocean water at a point close to being in equilibrium with limestone. Calculations show that this is near a pH of 8.0 or mildly alkaline. On the chemical pH scale, 7 is neutral. Acidic solutions have an excess of H^+ ions and a pH of less than 7. Alkaline solutions, with a pH greater than 7, have an excess of OH^- ions. Although the pH of ocean water can vary from about 7.5 to 8.5 depending on local controls, measurements show the long term average tends to be close to 8.0. [3]

Figure 62 illustrating how pH changes during titration experiments with increasing carbon dioxide in air shows the relation is logarithmic.[4] The graph shows that at first a small addition of carbon dioxide produces

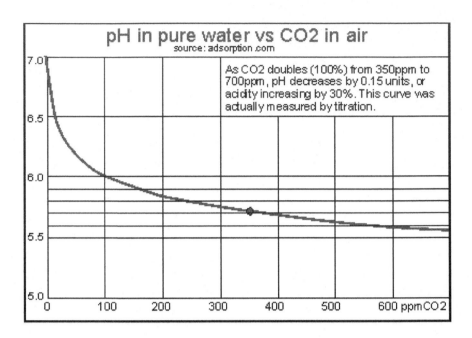

Figure 62. pH in water vs. carbon dioxide in air.

a large drop in pH, but as more and carbon dioxide accumulates in the water, less and less change in pH occurs. Doubling carbon dioxide from 350 to 700 ppmv lowers the pH only 0.15 pH unit. Another doubling to 1,400 would have even less effect on pH.

For seawater, a graph like this is misleading because of the buffering action of the real world situation. As ocean water absorbs carbon dioxide from the atmosphere, and the pH drops, the extra H+ ions react with calcium carbonate, releasing OH- ions; thus, this buffering reaction helps keep the pH near 8.0.

Finally, from this simplified discussion, it should be clear that the term "ocean acidification" is a misnomer chosen to shock the public. Ocean water with a pH of 8 is alkaline; so is ocean water with a pH of 7.9 or 7.5. Instead of reporting this basic fact, the sensationalist media makes it sound like fish will soon be swimming around in battery acid.

People are concerned about the short term changes. If it takes thousands of years for the buffering process to operate, it might too late for most shelled invertebrates. A change in ocean chemistry like this could produce a mass extinction event. Al Gore says this is already beginning.

Of course, and global warming alarmists argue that the pH of the oceans has already dropped about 0.1 pH unit, and that such a sharp drop like this has never happened before. If it continues, they say, it will produce dire consequences, which is exactly what they say will happen unless we take drastic action to reduce carbon dioxide emissions.

This raises the question of whether a short term drop in pH is abnormal. Research is available to help answer this question. A peer-reviewed study in 2005 used boron isotopic composition in 300-year old southwestern Pacific coral as a proxy for ocean water pH. They found that pH had varied since 1710 over a range of about 0.25 pH units, and that the change was cyclic following the 50-year Interdecadal Pacific Oscillation.[5] A study published this year also used boron isotopes to determine the pH history of corals from the South China Sea back to 7000 years BP. The scientists found that pH continually oscillated during this time over a range of about 0.4 pH units.[6] Their research showed the sharpest drop in pH in the entire record was from 6,500 to 6,000 years ago, and that pH did not correlate with the, "...atmospheric carbon dioxide concentration record from Antarctica ice cores...over the mid-late Holocene up to the Industrial Revolution."

Of course, the pH of the oceans varied in the past, and limestone sediments serve to buffer the process. The problem now, though, is that we're dumping carbon dioxide into the atmosphere at a very high rate. The oceans won't be able to change fast enough for life to adjust, and we already are seeing the evidence of this.

Let's take a look at whether the oceans are changing too fast for life to adjust. There are two ways to assess how realistic this scenario might be. The first is to look at earth history for what happened in the past, and the second is to assess how life might actually adjust to a lower pH.

It is important to remember from Chapter Two that in terms of earth history, what is unusual about the amount of carbon dioxide in the atmosphere today is that it is so low. For much of the last 550 million years, carbon dioxide levels in the atmosphere were several times higher than today, 10, 15, maybe even 25 times higher in the Cambrian. Certainly, if there is any element of truth to the scary story that ocean water pH will drop so much that invertebrates can't make shells or that shells will dissolve, we should see great gaps in the fossil record, times without invertebrate fossils, but we don't. Anyone who spends any time looking at outcrops of marine limestones from the Paleozoic Era, 550-250 million years ago, is inevitably struck by how many of these rocks are literally crammed full of fossil shells. If such animals prospered in ocean water under an atmosphere with significantly more carbon dioxide than we can ever reach today, how real can the scare about modern invertebrates really be? Geologic history tells us that this threat is greatly overblown.

I thought some huge mass extinctions occurred somewhere back then. Isn't that correct?

Geologists refer to the "Big Five," the five greatest mass extinction events. Each wiped out a significant proportion of all species living at the time. All five of them occurred during the Paleozoic and Mesozoic eras. There is no proof, however, that any of them is linked to an abrupt change in carbon dioxide. For instance, the extinction about 65 million years ago at the end of the Cretaceous Period is now generally accepted to have been caused by a mountain-sized asteroid or comet crashing into the earth.

Besides earth history, there is a lot of current research on how life might or might not be adjusting in less alkaline ocean water. A 2009 study published in *Proceedings of the National Academy of Sciences* found that a species of starfish thrived in warm water and an atmosphere containing 780 ppmv carbon dioxide, more than double today's level.[7] Another peer-reviewed study investigated how meiofauna, tiny animals such as nematodes, fared in an atmosphere containing more than six times the current amount of carbon dioxide. The results showed, "no significant difference in the abundance of total meiofauna, nematodes, harpacticoid copepods (including adults and copepodites) and nauplii by the end of the experiment," which lasted 56 days, and that, "elevated carbon dioxide had not impacted the reproduction of nematodes and harpacticoid copepods."[8]

A peer-reviewed study of sea urchins produced similar results. The researchers investigated the effects of pH as low as 7.6 and found no effect on egg fertilization for five different species; however, water temperature above 24 degrees C (75.2 degrees F) did affect development of one of the species.[9] The researchers speculated that this result might not have occurred if the gametes had been acclimated. Another recent article found a species of mussel doing quite nicely in water near a submarine volcano with a pH as low as 5.36, strongly acidic.[10] The scientists also noted that "many other associated species" lived in the same water.

Maybe lower ocean water pH doesn't damage all invertebrates equally, but coral bleaching is a fact. All over the world, coral is bleaching and dying. Al Gore says they are as important to ocean species as rain forests are to land species, and that global warming is rapidly killing them.

It is important to understand what coral bleaching is and what it is not because alarmists such as Al Gore imply that it is the same thing as death of the individual coral animal, the polyp, which is false. A coral reef is actually a community of organisms, especially certain species of algae living in a symbiotic (mutually beneficial) relationship with the individual coral polyps. The reef itself, composed of calcium carbonate, provides a place for the algae to live, and in return, the single-celled algae provide food for the coral as well as the bright colors associated with reefs. Changing environmental conditions can cause the algae to move to another area, leaving the gray to white calcium carbonate of the coral behind. This is referred to as coral bleaching, and it may be followed by death of the polyps, but that's often not the end result.

Yes, Al Gore says a lot about coral bleaching and the destruction of coral reefs, and a lot of it is a convenient fiction useful in achieving his political ends. As a 2003 study published in the journal *Science* makes clear, "Degradation of coral reef ecosystems began centuries ago... All reefs were substantially degraded long before outbreaks of coral disease and bleaching."[11]

Coral has a long history of forming reefs in the warm water of tropical seas, going back almost 500 million years. Two main forms were the dominant reef builders until they began to decline toward the end of the Paleozoic. They did not survive the mass extinction event 250 million years ago that separates the Paleozoic from the Mesozoic, but in the Triassic, the first geologic period of the Mesozoic Era, scleractinian corals appeared, the modern type, possibly as descendants of one of the earlier types. The oceans were considerably warmer at this time than today. Scleractinian corals became the most prominent type of reef builder by the end of the Triassic and on through the Jurassic, but declined in the Cretaceous only to expand again in the modern era, the Cenozoic.

This means that for most of the last 250 millions years, coral did just fine with warmer seas and an atmosphere containing far more carbon dioxide than today. Both of these facts suggest that coral should be able to adapt to the rather minor changes associated with the global warming theory. In fact, according to a peer-reviewed study published in 2004, warming that the IPCC projects for 2100 could increase coral reef growth by 35%.[12]

Research is showing us ways in which coral might adapt to changing environmental conditions. For example, we now understand that, instead of dying, coral may recover after bleaching by replacing one set of symbionts with another.

Replacement of symbionts is informally referred to as the "symbiont shuffling hypothesis," and seems to have been first suggested in 1993.[13] A 1997 peer-reviewed article said that, "coral communities may adjust to climate change by recombining their existing host and symbiont genetic diversities,"[14] and another in 1999 concluded that bleaching, "may be part of a mutualistic relationship on a larger temporal scale, wherein the identity of algal symbionts changes in response to a changing environment."[15] Several other studies have reached similar conclusions.

Extensive experimental and observational evidence supports the symbiont shuffling hypothesis. Only a couple of representative studies will be mentioned.

A 2004 article in *Science* describes an experiment in which scientists artificially bleached coral in the Caribbean by removing the algae. They then exposed the polyps to a variety of algae types and within six weeks, new symbiotic relationships had developed, but not always with the previous species.[16] These results suggest that the ability to change to a new symbiont partner may be a survival mechanism the coral has developed.

Another 2004 study produced observational evidence supporting these experimental results. Studying coral affected by bleaching following the strong 1997-98 El Niño-Southern Oscillation (ENSO) showed that coral containing heat tolerant algae became more abundant and more like coral from high temperature environments. The scientists concluded that, "these adaptive shifts will increase the resistance of these recovering reefs to future bleaching."[17]

Peer-reviewed studies supporting coral bleaching as a survival mechanism have continued each year since 2004. In one study year, Stanford scientists discovered healthy coral in a lagoon already as warm as the IPCC claims the earth will get by 2100. The coral hosted a warmth-adapted type of algae, which was rare on coral in nearby cooler lagoons. An illustrated description of this research is available on the web.[18] Another research project that reached similar conclusions was conducted near the southern end of the Great Barrier Reef where extensive coral bleaching resulted from very high water temperatures in the summer of 2006. Just over a year later, the coral were rapidly recovering suggesting a strong regenerative capability.[19]

In 2008, an article published in the journal *Coral Reefs* summarized the reef situation. The authors criticize the widespread use of sweeping predictions based on untested assumptions such as "all corals live close to their their thermal limits," and "corals cannot adapt/acclimatize to rapid rates of change." After reviewing the literature, the article concludes that, "it is premature to suggest that widespread reef collapse is a certain consequence of ongoing bleaching, or that this will inevitably lead to fisheries collapses."[20]

Obviously to say or imply that coral bleaching is, "usually a prelude to the death of the coral," as Al Gore does on p. 164 of his book, is to mislead. Corals are a whole lot more resilient than global warming alarmists give them credit for, but from a geological viewpoint, that is to be expected. After all, they survived one of the greatest mass extinction events the earth has ever experienced, the one at he end of the Cretaceous 65 million years ago that wiped out the dinosaurs.

What about polar bears? They're in such danger they've been placed on the endangered species list.

Well, actually they were listed as "threatened" in March 2008, not the same as listing them as endangered.[21] The great polar bear scare started a few years ago when large summer melting of arctic ice was occurring. A picture showing a polar bear and its cub floating precariously on a piece of ice surrounded by water became global warming alarmists' poster child. It was reproduced thousands of times, and just about everybody has seen it. Only later did Australian student Amanda Byrd reveal that she had taken the picture near an Alaskan shoreline while on a field trip and that the pair was not in danger.[22]

Scientists estimate that about 25,000 polar bears live in the wild, with 2/3 of the population in Canada.[23] Polar bears evolved about 200,000 years ago from the giant Alaskan brown bear, which means they have already survived two previous interglacials with warmer temperatures than today, as discussed in Chapter Four.

Global warming alarmists have succeeded in linking the fate of polar bears to Arctic sea ice. As Steven Amstrup of the United States Geological Survey says, "There is a definite link between changes in the sea ice and the welfare of polar bears. As the sea ice goes, so goes the polar bear.[24]" It seemed like a canny strategy a few years ago when ice in the Arctic ocean was decreasing, but now that it's increasing again (see Chapter Ten) alarmists could find it difficult arguing that sea ice is not so important after all. They face another difficulty, the results of the only recent census of polar bears.

The government of Canada's newest and largest province, Nunavit, commissioned polar bear expert Mitchell Taylor to undertake a count of polar bear numbers. His survey showed that their numbers had increased greatly from the previous count. For example, in the Davis Strait area, his survey showed 2,100, well over twice the 850 that were counted in the mid-1980s. "There aren't just a few more bears," Dr. Taylor said, "there are a hell of a lot more bears."[25] Dr. Andrew Derocher, of the World Conservation Union, still thinks polar bears are threatened because two of the thirteen bear populations in Canada are declining.[26] He attributes this to melting arctic ice. This seemed to be happening in 2007, when he made his view known. However, since then, as discussed in Chapter Ten, arctic sea ice has expanded by a half million square km.[27]

More ice in the arctic hasn't stopped the campaign to "save" polar bears. Preparations were under way as this was being written for the Copenhagen meeting in December 2009 where government representatives hope to pass the next climate change treaty. The chairman of the Polar Bear Specialist Group, Dr. Andy Derocher, a former student of Dr. Taylor, scheduled a preliminary meeting there this summer to plan their tactics for December. Dr. Taylor had already secured funding to attend when he received a message that his attendance had been rejected. In an explanatory email, the chairman said Dr. Taylor was not being invited because of his views on global warming which run, "counter to human-induced climate change are extremely unhelpful."[28] Dr. Derocher is right--Dr. Taylor's view would be unhelpful in generating more scare stories about polar bears in time for the December meeting.

SUMMARY

"Industrial carbon dioxide is turning the oceans acidic, threatening the foundation of sea life." So trumpets *Discover* magazine in its July 2008 issue. The scare is that shelled invertebrates won't be able to get enough calcium carbonate from ocean water to build their shells because changes "are happening much faster than animals can adapt."

Is this a real concern? Because the chemical reaction is reversible that produces carbonic acid when carbon dioxide dissolves in water (see Chapter Two) a buffering process in the oceans tends to keep their pH within a range of 7.5 to 8.5, which is alkaline rather than acidic. Alarmists accept this but say the pH has already dropped by 0.1 pH unit to 7.9, and that such rapid change proves buffering operates too sluggishly to prevent catastrophe.

Several factors argue against this. Boron isotope proxy studies show that rapid changes up to 0.4 pH units have occurred in the past without shelled invertebrate extinction. It should also be remembered that carbon dioxide in the atmosphere over much of the past 550 million years was much higher than today, 5-10 times higher, and even more, but the fossil record shows marine invertebrates remained abundant. There are also studies showing that certain modern invertebrates adapt successfully to higher carbon dioxide levels.

Alarmists say coral bleaching is an early example of the danger posed by warmer, less alkaline ocean water (they call it "acidification). Bleaching refers to the gray appearance coral has after the algae symbionts are lost. The coral looks dead, but appearances can be deceiving. The "symbiont shuffling hypothesis" is now well supported. This hypothesis holds that bleaching is a survival mechanism that allows coral to weather changing environmental conditions by replacing one set of algae symbionts with another. Recovery of bleached coral though such a replacement process has been observed in the Atlantic, Indian and Pacific oceans. Experimental evidence also supports the hypothesis.

Coral seems to be much more resilient than global warming alarmists imply, but that should not be a surprise. After all, coral did survive the massive extinction event 65-million years ago that killed the dinosaurs.

The same can be said for the poster child of the alarmists movement, polar bears. They lived through the warmer-than-today temperatures of the two previous interglacials and seem to be doing just fine now. A survey recently conducted in the largest Canadian province found the population up from 850 in the 1980s to 2,100 at the time of the count. This is significant because Canada is estimated to have 2/3 of the world's polar bears. It should be no surprise, therefore, that the polar bear expert who conducted this census was "disinvited" from attending a meeting of the Polar Bear Specialist Group in a run-up to the big Copenhagen meeting in December because his views are "extremely unhelpful." This comes in spite of the increasing Arctic ocean ice in recent years after alarmists had argued the bears were endangered because of decreasing ice.

NOTES AND SOURCES

(1) *Discover*, July 2008, p. 28, Kathleen McAuliffe, "Ocean Reflux."
(2) *An Inconvenient Truth*, Rodale Publishing Co., Emmaus, PA , p. 168, 2006, Al Gore
(3) An introductory discussion of oceanic pH is at this site: http://www.seafriends.org.nz/issues/global/acid.htm
(4) *Ibid.*
(5) *Science*, Vol. 309, p. 2204, 2005, Carles Pelejero et al, "Preindustrial to modern interdecadal variability in coral reef pH."
(6) *Geochimica et Cosmochimica Acta*, Vol 73, p. 1264, 2009, Y. Liu et al, "Instability of seawater pH in the South China Sea during the mid-late Holocene: Evidence from boron isotopic composition of corals."
(7) A summary of the article is available at the *New Scientist* site: http://www.newscientist.com/article/mg20227104.800-starfish-defy-climate-change-gloom.html#

(8) *Journal of Marine Science and Technology*, Vol. 15, p. 17, 2007, H. Kurihara et al, " Effects of elevated seawater carbon dioxide concentration of the meiofauna."

(9) *Proceedings of the Royal Society B*, Vol. 276, p. 1883, 2009, M. Byrne et al, "Temperature, but not pH, compromises sea urchin fertilization and early development under near-future climate change scenarios."

(10) *Nature Geoscience*, Vol. 10.1038NGE0500, 2009, V. Tunnicliffe et al, "Survival of mussels in extremely acidic waters on a submarine volcano."

(11) *Science*, Vol. 301, p. 955, 2003, J.M. Pandolfi et al, "Global trajectories of the long-term decline of coral reef systems."

(12) *Geophysical Research Letters*, 10.1029/2004/GL021541, 2004, B.I. McNeil, "Coral reef calcification and climate change: The effect of ocean warming."

(13) *Bioscience*, Vol. 43, p. 320, 1993, R.W. Buddemeier et al, "Coral bleaching as an adaptive mechanism."

(14) *Nature*, Vol. 388, p. 265, 1997, R. Rowan et al, "Landscape ecology of algal symbionts creates variation in episodes of coral bleaching."

(15) *American Zoologist*, Vol. 39, p. 80, 1999, R.A. Kinzie, "Sex, symbiosis and coral reef communities."

(16) *Science*, Vol. 304, p. 1490, 2004, C.L. Lewis & M.A. Coffroth, "The acquisition of exogenous algal symbionts by an octocoral after bleaching."

(17) *Nature*, Vol. 430, p. 741, 2004, A.C. Baker et al, "Corals' adaptive response to climate change."

(18) *Marine Ecology Progress Series*, Vol. 378, p. 93, 2009, T.A. Oliver & S.R. Palumbi, "Corals and starfish that defy climate change gloom." A summary of the article is posted here: http://theresilientearth.com/?q=content/heat-resistant-corals-ignore-climate-change-threats

(19) Recovery of the bleached reef is described at this site: http://www.climateshifts.org/?p=1426

(20) *Coral Reefs*, Vol. 27, p. 745, 2008, J.A. Maynard et al, "Revisiting the Cassandra syndrome; the changing climate of coral reef research."

(21) The federal rule listing polar bears as "threatened" is here: http://alaska.fws.gov/fisheries/mmm/polarbear/issues.htm

(22) *Daily Telegraph* UK, June 27, 2009, Christopher Booker, "Polar bear expert barred by global warmists." Video of the newscast announcing the story is at this site: http://www.abc.net.au/mediawatch/watch/default.htm?program=mediawatch&pres=20070402_2120&story=3

(23) *The Christian Science Monitor*, May 3, 2007, available online here: http://www.csmonitor.com/2007/0503/p13s01-wogi.html

(24) Quote from emagazine.com: http://www.emagazine.com/view/?3896

(25) *The Telegraph*, UK, March 9, 2007. The story is available on line here: http://www.telegraph.co.uk/news/worldnews/1545036/Polar-bears-thriving-as-the-Arctic-warms-up.html

(26) *Ibid*, #23.

(27) Arctic sea ice daily update: http://arctic-roos.org/observations/satellite-data/sea-ice/ice-area-and-extent-in arctic/?searchterm=Sea%20Ice%20Daily%20updated%20time%20series

(28) *Ibid*, #22

CHAPTER FOURTEEN—CONSENSUS AND SCIENCE

Isn't it true there is a consensus in the scientific community that climate change is the result of human activity?

That's partially correct. There is a consensus in the media, the government and among politicians, but not among scientists.

No, there is a consensus among scientists. Al Gore writes on p. 262 in *An Inconvenient Truth* that of the 928 "peer-reviewed articles dealing with 'climate change' published in scientific journals during the previous ten years," not a single one was "in doubt as to the cause of global warming." Isn't that correct?

This question refers to the 2004 article published in the journal *Science* by Naomi Oreskes, a historian at the University of California at San Diego.[1] She searched the ISI Web of Science Database over a ten-year period, 1994-2003, for the keywords "global climate change," and found 928 articles. Reading only the abstracts, she concluded that "none of the papers disagreed with the consensus position."

The answer to the question is that yes, such a study was published and yes, that is what the author concluded. That is not the same thing as saying a consensus exists among scientists regarding the causes of climate change.

Given the data cited in the article, how can that be true? What other interpretation can there be?

There are numerous potential flaws in such a study, not the least of which is the researcher's own personal bias. One tends to see what one wants or expects to see, unless the study methodology has safeguards to prevent this, but Dr. Oreskes' study included no such safeguards. If she assumed there was a consensus position, she might be expected to find that result, as indeed she did. A better-designed study might have used several people with differing viewpoints to rate the abstracts, but she did not. A different result might also have been obtained if the study included reading the text of the articles instead of only the abstracts, but again, that was not done. Because of such flaws, this widely cited study cannot be considered definitive.

Perhaps it was not a perfect study, but mere speculation is not enough to dismiss the results.

The same is true of Dr. Klaus-Martin Schulte's 2008 update of Dr. Oreskes' study. Using the same ISI database and the same methodology, he examined the abstracts of 528 articles on climate change published between 2004 and 2007. The results, published in the *Journal of Energy and Environment*, show that only 7% of the articles accept that humans cause global warming while 6% reject it outright.[2] The largest category of articles takes no position. Perhaps Al Gore will note that Dr. Schulte's newer study shows there is no consensus in the next edition of his book.

What about the IPCC's Fourth Assessment Report published in 2007? It gives a figure of "90% likely" that humans are causing climate change. And isn't there a statement that Nobel-Prize-winning scientists signed that humans cause global warming?

The Intergovernmental Panel on Climate Change, as part of the United Nations, can hardly be considered a non-political or non-governmental organization. In fact, many climate scientists criticize the IPCC for being overtly political, as political as Congress.

They say the same about the various scientific organizations and national Academies of Science that endorse the IPCC report. And, yes, as Al Gore says in *An Inconvenient Truth*, forty-eight past winners of a Nobel Prize, did sign a statement attesting to the dangers of global warming, but not a single one of them is a climate scientist. Just being a scientist, even one who won a Nobel Prize, does not make one an expert in all fields. A physicist has no special expertise in economics and a rocket scientist is not the expert one should go to for a heart bypass.

Although no one likes to admit it, since our society is considered so open-minded, the way government and the media treat the theory of global warming is exactly the sort of thing Galileo faced. Those in power told him with sincere belief and conviction that it was 100% certain the sun goes round the earth. Galileo insisted the opposite in his books, which recounted his own telescopic observations proving the consensus theory wrong. Surely, he thought, anyone making the same observations would come to the same conclusion. He even let one of the church leaders look through the telescope for himself. Galileo naively thought facts and data would be of greater importance than dearly held political and religious views. Nevertheless, his works were placed on the list of forbidden books, a convenient way to muzzle one's critics that is fortunately not currently available to global warming alarmists. Of course, in the end, Galileo was proven right, but it wasn't until 1992 that the Catholic Church issued a formal apology for his trial, which nearly cost him his life.

The only reason consensus is discussed at all in relation to global warming is that the theory has been moved from the realm of science into the political arena. Consensus finds no place in science. Science does not progress by consensus, popularity or government edict. Instead, facts and data are paramount. What is the data saying? What is the meaning? *All* the data, not just the facts that fit one's favorite notion, wishes or prejudices. Not how the world could be or should be, but how it actually is.

Consensus is brought up only when the science is shaky. Nobody says there's a consensus that the planets orbit the sun or that life is based on carbon molecules. But when the science isn't there, advocates try to have the game called by invoking consensus.

What is important in science is reproducible results. This is why science has been the most successful enterprise yet discovered for advancing the human condition. It's why people born now can expect to live into their eighties rather than their thirties, as was true just a few centuries ago.

Perhaps Michael Crichton summed up the unimportance of consensus in science best in a 2003 lecture at the California Institute of Technology when he said, "The work of science has nothing whatever to do with consensus. Consensus is the business of politics. Science, on the contrary, requires only one investigator who happens to be right."

Of course, science isn't decided by majority vote, but even scientists have to be practical. If not consensus, then how do they decide what's worth working on?

Observation, experiment, data and conjecture--that's what counts. Follow the data, go where it says to go, but scientists sometimes make mistakes. Interpretations can be shown to be wrong and technical developments can and do allow better data to be acquired. This is why scientific theories are never carved in granite. Unlike religious and political dogma, science is open to change.

Newtonian gravitation passed every test for two centuries and physicists thought it was infallible, until Einstein published his theory of special relativity in 1905 and a few years later, his general theory. By 1919, observable fact showed that Newton's law did not apply everywhere. To put it bluntly, it was wrong, and relativity replaced it, although it remained controversial for a long time. Now relativity has so far passed every test, but has it been proven true? No, just like any scientific theory, it has not been proven true. Someday Einstein, just like Newton before him, may be overturned. That's how science advances. Scientific progress would end if scientists behaved like global warming alarmists, yelling that the science is already decided and ignoring all contrary data.

Many scientists accept Sir Karl Popper's view that science progresses through falsification. That is, although a scientific theory, unlike a mathematical theory, can never be proven true no matter how much evidence is gathered, one single contrary fact can prove the theory wrong. It then must either be modified to encompass the new data or, if that's not feasible, abandoned completely. This process is the essence of the scientific method, and makes science a self-correcting activity.

A well-known quotation from Einstein illustrates this principle. In the years following the introduction of relativity, great opposition arose. Upon learning of one of the critical publication called *One Hundred Authors Against Einstein,* he said, "If I were wrong, then one would have been enough."

Falsification is reasonable as a philosophical principle. However, when presented with an urgent problem, one must be practical. If people see smoke coming from the attic of a house, should they call 911 or knock on the door and tell the owner of the house to climb into the attic to see if there are any flames?

Of course, there are times when action must be taken based on the preponderance of evidence. In reference to global warming, this means scientific evidence, not media hype or political posturing. Consider one well-known magazine, *Popular Science,* the September and October issues, 1977, Parts I and II of a lengthy article called "Colder Winters Ahead," illustrated by pictures of glaciers and cars buried in snow. The article dates from the end of a cooling period that started back in the 1940s. The consensus view in the media in late 1970s was that we were heading into an ice age. Newspapers and magazines published such articles regularly, including respected journals such as *National Geographic* (see the November 1976 article "What's Happening to our Climate").

Imagine what might have happened if politicians had acted on yesterday's consensus and taken action to try to do something about it. Luckily, the climate was already starting to turn in the other direction by late 1977.

Few would disagree that science considers new evidence when it becomes available. A lot of new data has been acquired since 1977. This is the reason a consensus has formed that the earth is warming dangerously, and not a consensus that exists only in the media, but a consensus among scientists. Isn't this correct?

Certainly, it exists in government, on television, National Public Radio, the Discovery Channel and in newspapers and magazines. For all of these, the science of global warming is settled, and there's nothing left to discuss. They treat global warming theory as a fact and allow no exposure of any study contrary to this view. The same is true of politicians and government funding agencies.

For example, the March 2009 issue of *Discover* on p. 14 carries a half-page advertisement announcing a conference jointly hosted by the magazine and the National Science Foundation called the Challenges of Climate Change. Included is the statement, "There is no longer any question that we are in the midst of an unprecedented era of rapid climate change." It is not a conference to discuss the merits of the question or even the strength of the evidence. The debate is over and the science is settled.

How can one argue with that? A respected science magazine and the chief government-funding agency for scientific research in the United States say it's a fact. How much chance might one have getting a grant from the NSF for a project that might cast doubt on this view? Sorry, out of luck. Go somewhere else for funding.

The problem is that to believe we live in an "unprecedented era of rapid climate change" requires pitching the entire science of geology out the window because any geologist can cite several examples of past periods of more rapid climate change. Chapters Three and Four discussed rates of climate change in greater depth.

The consensus today concerning global warming exists in the media and in the government, but not among climate scientists and those who study past climates. It's ironic that global warming alarmists make the same claim about the late 1970s; the only consensus then for a return to ice age conditions was in the media, not among scientists. Today, however, they argue it's different. Now the scientific evidence is irrefutable; they say the earth is in a period of unprecedented warming caused by humans. As Laurie David, the award-winning producer of *An Inconvenient Truth*, said, "No credible peer-reviewed scientist in the world disagrees any longer that the globe is warming and that humans are causing it."

In spite of such wildly exaggerated statements, many credible scientists remain unconvinced, scientists who publish articles in peer-reviewed journals.

Both sides make assertions in this debate. Does a consensus exist among scientists who study climate change and past climates or not? How is one to decide?

The most reliable way to answer this question is to go to the original sources, the scientific journals. Read the articles concerning global warming. This sounds reasonable, but for non-experts, it can be a daunting task. The articles are full of arcane language and technical expertise is required to understand many of them. A guide to relevant articles would be helpful. Fortunately, there are several such guides intended for lay people. A good place to begin is only a mouse click away:

http://petesplace-peter.blogspot.com/2008/04/peer-reviewed-articles-skeptical-of-man.html

This site lists more than 210 peer-reviewed articles with publication dates between 2000 and 2007. They are conveniently divided into nineteen categories such as Antarctica, Medieval Warming Period and Solar. A link is provided to each of the articles. For some of them, simply reading the titles makes it clear they are skeptical that humans cause global warming. Consider for example, "Geophysical, archaeological, and historical evidence support a solar-output model for climate change," and "Gulf Stream safe if wind blows and earth turns."

The previous web site is a good beginning place for the scientific literature, but a much larger web-based database of the scientific literature concerning climate is maintained and regularly updated by the Center for the Study of Carbon Dioxide and Global Change. The center is mainly the work of brothers, Craig and Keith Idso, sons of the well-known global warming skeptic Sherwood Idso. The link to the database is here: http://www.co2science.org/subject/subject.php

The number of articles referenced in this database is truly huge and growing. Although global warming skeptics maintain the database, articles accepting the so-called consensus position are also included. Clicking on a link to a particular article leads to a short summary consistently presented in the same format: Reference; what was done; what was learned and what it means. This is especially useful because a large number of articles are available in this database, more than most people would want to read. Because each listing includes the complete reference to the journal where the article was published, any skepticism about whether the summaries are biased can be dispelled by visiting a university library and reading some of the articles at random.

Doing so will be an illuminating exercise for anyone who thinks the theory of global warming is settled science.

Sherwood Idso and his sons are rumored to be the recipients of funding from the Western Fuels Association, which in turn is

funded by coal and utility companies. With such an obvious bias, why should anyone believe a single word on their web site?

Ah yes, the ad hominem argument fallacy. When the argument is sound, attack the person. This widely used technique of propaganda is all too prevalent in negative ads that flood the airways during political campaigns. It's one of the most prevalent techniques global warming alarmists use against those who prefer to interpret the evidence for themselves rather than blindly accepting what others tell them. Such an attack carries the implication that one's work cannot be trusted because it will, out of financial necessity, reflect the viewpoint of the person or organization providing the funding. Even worse, it is meant to discredit a whole class of people, scientists skeptical of global warming.

When global warming alarmists raise this charge, why is the assumption always made that bias is only a one-way street? A scientist's work is tainted if funding comes from an energy company, but climate alarmists are given an automatic pass as being bias-free, whether their funding comes from the government or an environmental organization.

Scientific research, as it's practiced today, is mainly government science. The GAO reported to Congress in 2005 that fourteen different federal agencies spent $5.09 billion on climate research during the previous budget year.[3] That was during the Bush administration. The Obama administration is boosting this by billions more. Thirteen of the agencies involved in climate research, including the NSF, the Department of Energy and NASA, parcel out money to individual researchers through grants.[4] In order to receive funding, an individual researcher's proposed project must be favorably reviewed by an agency's review committee and pass muster with the upper-level bureaucrats. Given the current bias toward humans causing global warming, a proposal concerned with dire consequences and the need for government action is far more likely to receive funding than one that says there is no problem, or that the problem is minor.

This situation is not necessarily due to the individual scientists within a particular agency having personal bias against supporting such research, although certainly some feel that way. It's more about the agency looking out for its own best interests. Congress is far more likely to increase the funding, or at least not cut it, if an agency makes a persuasive case that grave problems face the nation and the agency's budget needs to be increased so it can more effectively work to solve them.

An interesting example illustrating of this situation recently came to light. Dr. John S. Theon retired in 2008 from his position in NASA as

Chief of the Climate Processing Research Program. Before that, he was Chief of the Atmospheric Dynamics and Radiation Branch. For years, NASA has been the most vocal among federal agencies beating the drum that action must be taken to prevent future disasters from global warming. James Hansen, head of the NASA Goddard Institute for Space Studies, has been at the forefront of this effort. Testifying to Congress on several occasions, Dr. Hansen, whose Ph.D. is in physics, not climatology, is second only to Al Gore as a global warming alarmist. His public pronouncements have grown particularly inflammatory. For example, in 2007, he compared coal trains to Nazi "death trains" on their way to crematoria.[5] During Dr. Theon's tenure, NASA doled out huge piles of money for research projects supporting this position. Although he was once Dr. Hansen's boss, now that he's retired and free to speak his mind, Dr. Theon announced in a January 15, 2009 letter to the House of Representatives Public Works Committee that he is proud to add his name to those "who disagree that global warming is man-made."[6] He goes on to call computer climate models "useless," and says "some scientists have manipulated the observed data to justify their model results." Of Hansen, Dr. Theon says he "embarrassed NASA," with his claims of cataclysmic global warming.

Because of the huge government funding discrepancy in favor of global warming alarmists, something on the order of 1000:1,[7] scientists who don't toe the party line must seek funding elsewhere, or give up their work. There are only two possibilities: environmental organizations and private industries. The environmental door is obviously barred, which leaves only private industry. Yet in the eyes of the media, accepting such funding makes their work worthless. On the other hand, scientists funded by environmental organizations are untainted. This seems especially naive. What would happen to environmental organizations involved in climate research if public and government opinion should turn against accepting global warming alarmism? Contributions supporting them would collapse and the organizations would soon cease to exist. Isn't an organization's survival a much greater incentive for bias than any conceivable loss energy companies might suffer if catastrophic global warming should be proven correct?

None of this applies to Al Gore. He's been crusading against global warming since he was a little known junior senator and he doesn't have to apply for research grants. It is true, is it not, that his is work is not tainted by financial interests?

Al Gore is especially interesting, but one must remember that he is a politician, not a scientist. As a winner of the Nobel Peace Prize in 2007, his motives must be pure, at least according to global warming alarmists. The following is presented not as an ad hominem attack on Gore, but as an illustration that the position alarmists take may not be untainted by financial interests.

When he conceded defeat to George Bush in January 2001, ABC News estimated in 2007 that his net worth was approximately $1 million. At the time of the broadcast, they estimated it at $100 million, a ten-thousand-percent increase in just six years.[8] How did he do it? In simplest terms, he turbocharged his war on global warming. There is, of course, his movie, and the book based on the movie and all his speaking fees. These are lucrative and well publicized, but the ABC story failed to mention his main business interest. He is the founder and chair of a London-based private equity firm, Generation Investment Management. The company appears to have considerable influence over the only company in the U.S. that trades carbon credits, Chicago Climate Exchange, CCX, something European countries are already doing.[9] So far carbon trading is voluntary in the U.S.; but, if Congress passes a law mandating it, as the House of Representative has already done, his company stands ready to cash in on it in a big way. It is not inconceivable that Gore might one day be the richest man in the world.

Are there any well-known environmentalists not in agreement with the consensus view?

Perhaps the best known is Lawrence Solomon, the founder of several environmental organizations including Friends of the earth Canada and the World Rainforest Movement. A prominent critic of energy companies, his 1978 book *The Conserver Solution*, became the bible of the environmental movement at that time. The one time advisor to President Carter became interested in why any real scientist would dissent from the "settled science" of global warming. He started to look into it, expecting the few he would find would be either in the pay of energy companies or kooks. To his surprise, he found a large number, many leaders in their fields, whose credibility is beyond question. He wrote a book detailing his experience, *The Deniers: Scientists Who Stood Up Against Global Warming Hysteria*.

In a 2008 interview with The Frontier Center for Public Policy in Winnipeg, he said, "Many scientists believe that CO_2 is a problem. Many others, perhaps the majority of scientists, believe that CO_2 is not a problem. Many scientists believe that CO_2 is actually a benefit. It wasn't actually that long ago that CO_2 was universally regarded as beneficial, as plant food." In response to other questions, he said, "Politicians don't realize that the science is not settled on climate change. They think it's a done deal. To date we have not had compelling evidence that climate change is either man-made or harmful."[10]

SUMMARY

Global warming alarmists consider the consensus argument to be one of their really big guns, one which skeptics have no defense against. The argument runs something like this:

The science of global warming is settled. What's important now is to get on with doing something about climate change before it's too late, not to sit twiddling our thumbs and wasting more precious time. We've already spent years letting a few loud-mouthed skeptics have their say. The media gave them equal coverage, they were treated fairly, and what came of it? Nothing, except the same old tired rhetoric, endlessly presented. Frankly, their case is pathetic as all the top climate scientists in the world recognize. To keep giving them media coverage would be like pretending we take seriously the arguments of kooks who still think the earth is flat. They've had their time, but now it's over.

To counter this debate-ending argument, this chapter discussed data indicating that only among activists, and their supporters in the media and government, does a consensus exist that human industrial activity causes the climate to warm. No such consensus exists among scientists who actually study past climates and climate change. Even if a consensus did exist among these scientists, it would be irrelevant in deciding the issue because consensus finds no home in science. The point is made that observations, experiments, facts and logically drawn conjecture drive progress in science, not wishes, hopes, opinions, prejudices or popularity. As Michael Crichton wrote in the *Wall Street Journal,* "If it's consensus, it isn't science. If it's science, it isn't consensus. Period."[11]

Certainly science has taken wrong turns in the past and will do so again, but it eventually finds its way onto the right road because it has a built-in self-correcting mechanism, the scientific method. Attempting to silence critics never finds a place at the table of science because no matter how many positive tests a theory passes, it can never be proven true. One failed prediction, however, one stubborn fact that a theory is unable to encompass, is enough to doom it. As T.H. Huxley, popularly known as "Darwin's Bulldog," said, "The tragedy of beautiful theories is that they are often destroyed by ugly facts."

NOTES AND SOURCES

(1) *Science*, Vol. 306, no. 5702, p. 1686, 3 Dec. 2004, Naomi Oreskes, "The Scientific Consensus on Climate Change."

(2) *Energy & Environment*, Vol. 19, no. 2, pp. 281-286, March 2008, Klaus-Martin Schulte, "Scientific Consensus on Climate Change?"

(3) GAO Report to Congressional Requesters, August 2005, Climate Change; Federal Reports on Climate Change Funding Should be Clearer and More Complete: http://www.gao.gov/new.items/d05461.pdf

(4) CCSP Annual Report to Congress: *Our Changing Planet* CCSP-8, 2008

(5) Dr. Hansen made his infamous coal-crematoria remark October 22, 2007 during testimony to the Iowa utilities board. The text of the relevant section is available here: http://dotearth.blogs.nytimes.com/2007/11/26/holocausts/
For those interested, the complete text of his remarks can be read at the following link: http://www.columbia.edu/~jeh1/2007/IowaCoal_20071105.pdf

(6) Excerpts from Dr. Theon's letter are posted here: http://epw.senate.gov/public/index.cfm?FuseAction=Minority.Blogs&ContentRecord_i d=1a5e6e32-802a-23ad-40ed-ecd53cd3d320

(7) This estimate is given in Chapter 2 of *Skeptics Guide to Global Warming* (2007) by Warren Meyer, published by Lulu.com. The text is available at this link: http://www.climate- skeptic.com/2007/09/chapter-2-skept.html

(8) The text from *Good Morning America* is given at this link: http://abcnews.go.com/GMA/story?id=3281925

(9) Deborah Corey Barnes discusses Gore's business activities in "The Money and Connections Behind Al Gore's Carbon Crusade," available here: http://www.humanevents.com/article.php?id=22663

(10) The entire interview with Lawrence Solomon is available at this link: http://www.fcpp.org/main/publication_detail.php?PubID=22681 1.The Wall Street Journal, November 7, 2008, printed Michael Crichton's January 17, 2003 lecture at the California Institute of Technology. His remarks can be read here: http://online.wsj.com/article/SB122603134258207975.html

ADDENDUM

The nature of the book publishing business in the United States makes it almost impossible to have a book available in print in less than two years from the time it was written. For most books, this matters little since most subjects don't change much in such a short span of time. For some fast-moving subjects, however, significant change can occur in two years. Global warming and climate change is one such subject. For that reason, the authors deemed it important to bring our discussion up to date just before the book goes to press, hence this addendum.

Although other authors might not agree, we have narrowed down our list of what we consider to be the most important developments in the field of global warming and climate change to three:

* The so-called Climategate Scandal

* The CERN cloud experiment

* Sunspot Cycle 23 and 24

Time allows us only to present brief summaries of each topic; however, we list references where more comprehensive accounts can be found. As throughout this book, we attempt to make our summations neutral.

CLIMATEGATE

Only a cursory internet search will find numerous accounts of the Climategate episode which came to light in mid November 2009, only a few weeks before the opening of the failed Copenhagen climate summit. Finding an account on the internet that is not biased to one side or the other in the climate debate is a much more daunting task. One can often detect the bias simply by looking at the headlines or introductory sentences of an account. Web sites run by global warming skeptics use terms such as, "whistle blowers, leaked emails and distorting data." In contrast, climate alarmist web sites mention "stolen emails, hacked email accounts and efficiently analyzing data."

As always, it is up to the individual to decide on the veracity of whatever might be read.

The essential facts of what global warming skeptics call the Climategate Scandal are that on November 20, 2009, an unknown person or persons released to several web sites 61 megabytes of compressed emails, data, files and computer code from a backup computer server at the Climatic Research Unit (CRU) of the UK's University of East Anglia. Included in this were more than a thousand emails from climate scientists spanning many years. Skeptics say the data proves that alarmist scientists intentionally manipulated climate data to make it support their view. In the case of the "hockey stick" graph (see Chapter Eight) they charge that data was either manufactured or falsified. In stark contrast, alarmists say they some of the emails show the scientists behaving badly or even like jerks but that most of them show them simply having frank discussions among their colleagues and peers.

One of the best known of the emails was from Professor "Phil" Jones at the CRU to Michael Mann, Raymond Bradley and Malcolm Hughes, the authors of the original "hockey stick" article of 1998. In this email, Professor Jones said, "I've just completed Mike's Nature trick of adding in the real temps to each series for the last 20 years (i.e., from 1981 onwards) and from 1961 for Keith's to hide the decline." Nature in this message refers to the international journal of science by that name. Skeptics says the email refers to intentional hiding or distorting temperature. Alarmists argue that the message instead simply refers to a method published in *Nature* for combining tree ring data with instrumental data. In that context, "decline" refers to the fact that recent tree ring data shows lower global temperatures than instrumental data.

Which side is right, or at least closer to being right? It is true that this email refers to a method of combining the two data sets. This may not be as innocent as it sounds, however, because most of the original hockey stick article showing no medieval warming is based on tree ring data being indicative of warming temperatures. This is an over simplification, however. Tree rings become wider not just because of warmer temperatures, but also increased rainfall and more atmospheric carbon dioxide. Carbon dioxide is especially important because the amount in the atmosphere today is 40% higher than a couple of hundred years ago. Remove the suspect tree ring data, and the original hockey stick article clearly shows a warmer Middle Ages than today.

There are other well-known innocent or damning extractions from the emails, depending on one's point of view. However, this brief addendum lacks room to include them. Use the references at the end of this postscript to find them.

One interesting aspect of the Climategate controversy that is rarely mentioned is related to the Fortran computer code Jones and the CRU used in determined their global temperature series, included in the released 61 MB of data and emails. The programmers included numerous remarks indicating that a lot of filtering of raw data occurred. For example, some of the code removed proxy data that did not correlate well with other data, or replaced proxy data with measured data. One of the remarks, in referring to a data set, said, "These will be artificially adjusted to look closer to the real temperatures."

It does seem that Mann and Jones should have at least mentioned this in their published articles.

THE CERN CLOUD EXPERIMENT AND SUNSPOTS

Possible influence of solar variations on earth's climate is discussed in Chapter Eight where it was pointed out that a strong correlation exists between solar activity (sunspots) and global climate. During a sunspot minimum (less active sun) the earth cools. Climate alarmists say the resulting variation in solar energy is so small any effect on climate would be minor without a positive feedback mechanism. Several years ago, Henrik Svensmark introduced a theory involving cosmic rays that provides just such a feedback. He postulates that a less active sun produces decreased cosmic ray shielding for the earth, increasing the cosmic ray flux, so that more clouds form. The clouds reflect more of the sun's energy into space thereby amplifying the cooling effect from lower solar irradiance.

In August 2011, the European Laboratory for Particle Physics (CERN) announced the first results of an experiment using the Large Hadron Collider designed to test Svensmark's theory. Although more experiments are required, these early results support the theory. Commenting on the findings, Svensmark said, "Of course, there are many things to explore, but I think that the cosmic-ray/clouding seeding hypothesis is converging with reality." CERN said in announcing the results, "We've found that cosmic rays significantly enhance the formation of aerosol particles in the mid troposphere and above. These aerosols can eventually grow into the seeds for clouds."

If the theory continues to survive additional testing, the underlying mechanism accounting for the relationship between sun spots and climate might finally be explained.

Chapter Eight also discussed the unusually long wait for sunspot cycle 24 to begin. From hindsight, it is now clear that Cycle 24 was beginning as the chapter was being written. It is a weak cycle that NASA predicts will peak in early 2013 with about half the number of sunspots as the previous cycle. If correct, this will be the weakest sunspot cycle in over eighty years. Cycle 24 differs in other ways too. A meeting of 320 top solar physicists in June 2011 discussed the current unusual behavior of the sun and whether the sun is entering another minimum. Frank Hill of the National Solar Observatory said that Cycle 25, "may not actually happen." As Chapter Eight stated, no one knows how a sunspot minimum will affect climate, but finding out should be interesting.

SUNSPOT CYCLE 23 AND 24

At the time the main text was being written, the sun appeared to be headed for "... its deepest sunspot minimum since the nineteenth century." How, in fact, did this turn out?

It was a remarkably accurate prediction with the bottom part of the cycle, reached in early 2009, having the least number of sunspots in many decades. The following maximum, peaking in late 2011-early 2012, also had the lowest number of sunspots in many decades. Many scientists think we may be entering a new solar minimum that could last for decades as the most recent maximum did. It will likely take a couple of decades to know for certain. As discussed in Chapter Eight, previous solar minimums were associated with times of global cooling.

This seemed to be happening again. Although carbon dioxide continued to increase in the atmosphere, the slow, jerky warming of the Earth stopped or paused from 1998 until 2014. In April 2015, a study reported that the main portion of the Antarctic ice had expanded to a record thickness, and a "surprisingly high" amount of geothermal heating was discovered beneath the West Antarctic Ice Sheet, so instead of warming air, it's warming rock affecting the ice in this area.

Alarmists initially said the lack of warming over this 15-year period was real; then, when it could no longer be explained away, they said it was simply due to "natural variation." One can't help but wonder why any warming is never ascribed to natural variation.

Now, in late 2015, a pronounced El Niño is in full swing, bringing warming temperatures. Of course, alarmists and the media will say global warming is the cause. They will also forget to mention that La Nina, bringing cooling, usually follows an El Niño event.

What of the future? Will the Earth warm? Will it cool? Will Democrats and Republicans continue their shouting, neither never hearing what the other says?

As a geologist, my money is on the Earth continuing pretty much as it always has for more than four billion years, to continue to change, gradually for the most part, but sometimes violently.

NOTES AND SOURCES

Climategate
http://en.wikipedia.org/wiki/Climatic_Research_Unit_email_controversy
http://factreal.wordpress.com/2009/11/23/climategate-global-warming-email-scandal/
http://www.factcheck.org/2009/12/climategate/
http://tech.mit.edu/V129/N57/davidson.html
http://wattsupwiththat.com/2009/12/01/lord-moncktons-summary-of-climategate-and-its-issues/

CERN Cloud Experiment
http://www.thegwpf.org/the-observatory/3702-cern-finds-qsignificantq-cosmic-ray-cloud-effect.html
http://solarscience.msfc.nasa.gov/predict.shtml
http://physicsworld.com/cws/article/news/46953
http://www.skyandtelescope.com/news/123844859.html

Sunspot Cyle 23 and 24

http://www.solen.info/solar/index.html
http://www.solen.info/solar/images/cycles23_24.png
http://www.insurancejournal.com/news/national/2015/10/15/385147.htm
http://www.weather.com/news/climate/news/el-nino-outlook-strong-possible-may2015

CPSIA information can be obtained at www.ICGtesting.com
Printed in the USA
BVOW06s0334131016

464923BV00002B/9/P

9 781937 327033